Electromagnetic Interference and Compatibility

Electromagnetic Interference and Compatibility

Editor

Paolo Stefano Crovetti

MDPI • Basel • Beijing • Wuhan • Barcelona • Belgrade • Manchester • Tokyo • Cluj • Tianjin

Editor
Paolo Stefano Crovetti
Politecnico di Torino
Italy

Editorial Office
MDPI
St. Alban-Anlage 66
4052 Basel, Switzerland

This is a reprint of articles from the Special Issue published online in the open access journal *Electronics* (ISSN 2079-9292) (available at: https://www.mdpi.com/journal/electronics/special_issues/eimc#).

For citation purposes, cite each article independently as indicated on the article page online and as indicated below:

LastName, A.A.; LastName, B.B.; LastName, C.C. Article Title. *Journal Name* **Year**, *Volume Number*, Page Range.

ISBN 978-3-0365-0500-8 (Hbk)
ISBN 978-3-0365-0501-5 (PDF)

Cover image courtesy of Paolo Crovetti.

Contents

About the Editor

Paolo Stefano Crovetti was born in Turin, Italy, in 1976. He received the Laurea degree (summa cum laude) and the Ph.D. degree in electronic engineering from the Politecnico di Turin, Turin, in 2000 and 2003, respectively. He is currently an Associate Professor of Electrical Engineering with the Department of Electronics and Telecommunications (DET), Politecnico di Torino, Turin. He has co-authored more than 75 articles appearing in journals and international conference proceedings. His research interests are in the fields of analog, mixed-signal, and power integrated circuit (IC) design and IC-level and System level Electromagnetic Compatibility (EMC). Prof. Crovetti is a Senior Member of the IEEE and serves as the Subject Editor-in-Chief of IET Electronics Letters in the area of Circuits and Systems and as an Associate Editor of the IEEE Transactions on VLSI Systems.

Preface to "Electromagnetic Interference and Compatibility"

Due to the continuous progress in semiconductor technology and the rapidly evolving application scenarios, electromagnetic compatibility (EMC) is constantly raising new challenges and is a very dynamic field of research. In this context, this collection of papers, originally published in the "Electromagnetic Interference and Compatibility" Special Issue of *Electronics* for which I served as a Guest Editor, offers a vivid picture of the EMC research challenges and directions over the last years in this complex and multifaceted field.

Focusing on EMC in communication systems, the paper "Interference of Spread-Spectrum EMI and Digital Data Links under Narrowband Resonant Coupling" by Crovetti and Musolino highlights how traditional methods like spread-spectrum clock modulation, developed with reference to AM and FM radio receivers, are no longer well suited to digital communications.

In the very crucial field of EMC in power electronics, the paper "Signal Transformations for Analysis of Supraharmonic EMI Caused by Switched-Mode Power Supplies" by Sandrolini and Mariscotti explores advanced signal processing techniques (Wavelet Packet Transform and the Empirical Mode Decomposition) in the analysis of electromagnetic emissions of power converters, while in "Modeling and Optimization of Impedance Balancing Technique for Common Mode Noise Attenuation in DC-DC Boost Converters," by Shuaitao Zhang et al. more conventional balancing techniques are optimized to attenuate common-mode emissions.

For new application scenarios, EMC challenges in emerging electric vehicles are addressed in "Electromagnetic Susceptibility of Battery Management Systems' ICs for Electric Vehicles: Experimental Study," by Aiello. Moreover, IC-level EMC issues in operational amplifiers are addressed in "EMI Susceptibility of the Output Pin in CMOS Amplifiers" by Richelli, Colalongo, and Kovacs-Vajna, and the susceptibility of Hall Effect sensors is studied in "Hall-Effect Current Sensors Susceptibility to EMI: Experimental Study" by Aiello.

New contributions in the area of cross-talk reduction and signal/power integrity are presented in "A Novel Meander Split Power/Ground Plane Reducing Crosstalk of Traces Crossing Over" by Jung-Han Lee and in "A Dual-Perforation Electromagnetic Bandgap Structure for Parallel-Plate Noise Suppression in Thin and Low-Cost Printed Circuit Boards" by Myunghoi Kim.

The active research area on the EMC properties of materials and their application to the suppression of interference is represented in this volume by "Shielding Properties of Cement Composites Filled with Commercial Biochar" by Yasir et al. and by "Synthesis and Characterization of Polyaniline-Based Composites for Electromagnetic Compatibility of Electronic Devices" by Gareev et al. Last, but not least, new contributions on ferrite cores and their characterization are presented in "Performance Study of Split Ferrite Cores Designed for EMI Suppression on Cables" by Suarez et al. and in "Simple Setup for Measuring the Response to Differential Mode Noise of Common Mode Chokes" by González-Vizuete et al.

Though not exhaustive, the papers collected in this volume can be useful to address practical EMC problems and stimulate future research and should be well received by the EMC community.

Paolo Stefano Crovetti
Editor

Article

Interference of Spread-Spectrum EMI and Digital Data Links under Narrowband Resonant Coupling

Paolo Stefano Crovetti *,† and Francesco Musolino †

Department of Electronics and Telecommunications (DET), Politecnico di Torino, 10129 Turin, Italy; francesco.musolino@polito.it

* Correspondence: paolo.crovetti@polito.it; Tel.: +39-011-090-4220

† The authors contributed equally to this work.

Received: 6 November 2019; Accepted: 17 December 2019; Published: 1 January 2020

Abstract: In this paper, the effects of electromagnetic interference (EMI) coupled to a radio-frequency (RF) communication channel by resonant mechanisms are investigated and described in the framework of Shannon information theory in terms of an equivalent channel capacity loss so that to analyze and compare the effects of non-modulated and random Spread Spectrum (SS) modulated EMI. The analysis reveals a higher EMI-induced capacity loss for SS-modulated compared to non modulated EMI under practical values of the quality factor Q, while a modest improvement in the worst case capacity loss is observed only for impractical values of Q. Simulations on a 4-quadrature amplitude modulation (4-QAM) digital link featuring Turbo coding under EMI resonant coupling reveal that SS-modulated EMI gives rise to higher bit error rate (BER) at lower EMI power compared non-modulated EMI in the presence of resonant coupling for practical values of Q, thus suggesting a worse interfering potential of SS-modulated EMI.

Keywords: Spread Spectrum; DC DC power converters; digital communications; channel capacity; resonant coupling

1. Introduction

Due to the widespread diffusion of highly integrated information and communication technology (ICT) systems like smartphones, tablets and intelligent sensor nodes, digital communication modules operate in a more and more harsh electromagnetic environment, in which electromagnetic interference (EMI) from digital integrated circuits (ICs) and switching mode power converters can be easily coupled to the nominal signal path via the silicon substrate, the IC package and the power distribution network of ICs and printed circuit boards (PCBs), resulting in a degraded bit error rate (BER) and possibly in a complete communication failure [1–7].

As a consequence, the effects of EMI need to be addressed adopting specific countermeasures as early as possible in the design of circuits and systems and/or by limiting the coupling mechanisms so that to assure the proper operation with no penalty in terms of costs, time-to-market and performance, and also in order to meet the stringent Electromagnetic Compatibility (EMC) regulations [8–11].

For this purpose, several approaches have been proposed in recent years to mitigate EMI [12–16]. Among them, *Spread-Spectrum (SS) techniques*, which consist in the modulation of the frequency of switching signals around the nominal value, so that to spread their spectral energy over a wider bandwidth [17,18], are often adopted in clock oscillators [19–23] and DC–DC converter controllers [24–27] since they give rise to a significant (10–20 dB) reduction of EMI spectral peaks measured following EMC standards [8–11] and make it possible to meet EMC requirements at low cost. In this field, chaos-based random SS-modulations have proven to achieve the best spectral characteristics [28–30].

While SS techniques help in complying with EMC regulations, their effectiveness in reducing the interfering potential of switching signals is more controversial and has been sometimes put in question [31–34]. Moreover, SS-modulations have found to bring no BER reduction in an I^2C link operating in the presence of EMI generated by a power converter [35] and a worse interfering potential of SS-modulated compared to non-modulated EMI in Wireless Local Area Networks (WLAN) is reported in [36].

In this scenario, focusing on the effects of SS EMI on digital communications, the impact of random SS-modulated interference on baseband digital data lines has been described in [37] in terms of an equivalent channel capacity loss and compared with non-SS-modulated EMI in practical application scenarios, revealing that SS-modulated EMI can show a worse interfering potential compared to non-modulated EMI in a digital link featuring advanced channel coding. The analysis presented in [37], however, was carried out assuming frequency-independent EMI coupling in the victim channel bandwidth. This assumption conveniently describes low frequency inductive and capacitive coupling scenarios in baseband data links but could not be valid in general in the presence of radiated and conducted EMI resonant coupling mechanisms [38–42], which affect high-frequency EMI propagation and can be therefore relevant when interference with radio-frequency (RF) digital communications are considered.

Generally speaking, in the presence of resonant coupling, the adoption of SS-modulations may lead to a more significant reduction of the EMI power which is actually coupled to the victim equipment compared with the case of wideband coupling considered in [37], as illustrated in Figure 1, where B and W are the channel and the EMI bandwidth, respectively. From the figure, it can be observed that under the same resonant coupling mechanism $G(f)$, the coupled EMI power can be reduced of more than 50% by adopting SS-modulations compared to the non-SS-modulated case. In this paper the analysis in [37] is extended to investigate if and to what extent the conclusions on the worse EMI-induced capacity loss of SS-modulated signals presented in [37] apply to the case of resonant coupling.

The paper is organized as follows: in Section 2, the model of a digital communication channel under EMI resonant coupling is introduced. With reference to such a model, a description of the adverse effects of EMI in terms of capacity loss is introduced in Section 3 and discussed in Section 4. Simulations on the effects of resonant-coupled EMI from a switching mode DC—DC converter on a 4-quadrature amplitude modulation (4-QAM) data link featuring Turbo coding are then presented in Section 5 to verify the theoretical results with reference to a practical case. Finally, in Section 6, some concluding remarks are drawn.

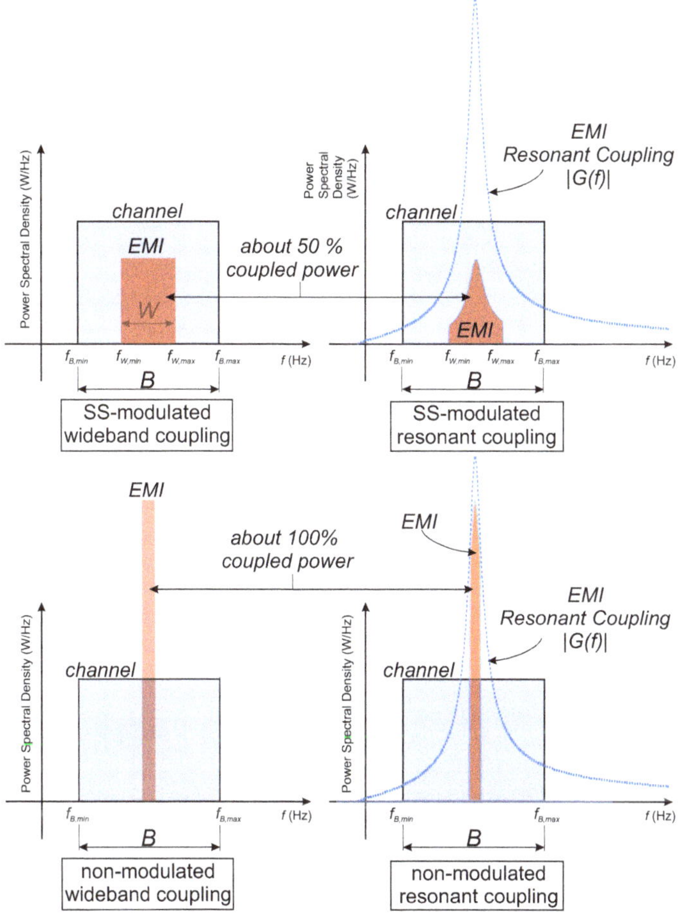

Figure 1. Effects of resonant coupling mechanisms on non-modulated and SS-modulated EMI.

2. Communication Channel Modeling under Resonant EMI Coupling

In this paper, the interfering potential of periodic and SS-modulated EMI generated by switching signals and coupled to a digital communication channel by a resonant coupling mechanism, as depicted in Figure 2, are compared in the framework of Shannon information theory in terms of an equivalent EMI-induced channel capacity loss [43], following the approach adopted in [37].

2.1. Channel Modelling

A band-limited communication channel corrupted by additive white Gaussian noise (AWGN) with unilateral power spectral density N_0 is considered in what follows. The transmitted signal and the background noise are both described by two wide-sense stationary (WSS), statistically independent gaussian random processes. In particular, the signal process is assumed to have a total power P_S and a power spectral density (p.s.d.) P_S/B constant over the channel bandwidth $B = \left(f_{B_{\min}}, f_{B_{\max}}\right)$, as depicted in Figure 3, while the noise is assumed to have unilateral power spectral density N_0.

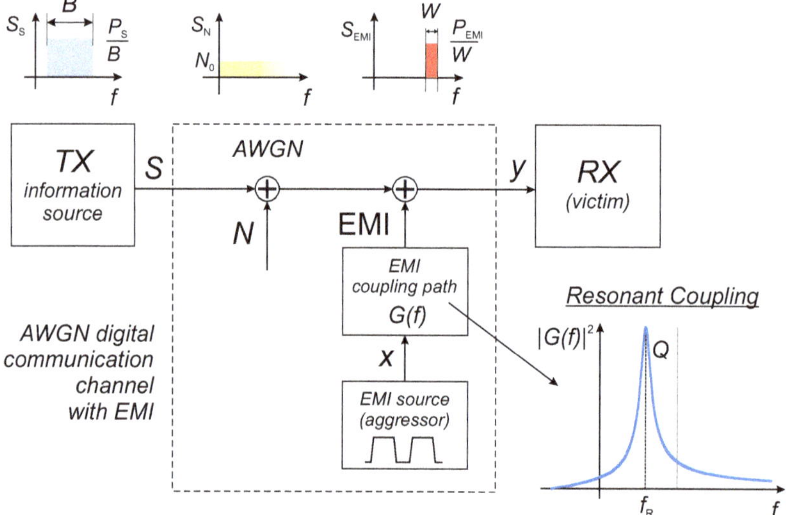

Figure 2. Digital communication channel with EMI resonant coupling: Block Diagram.

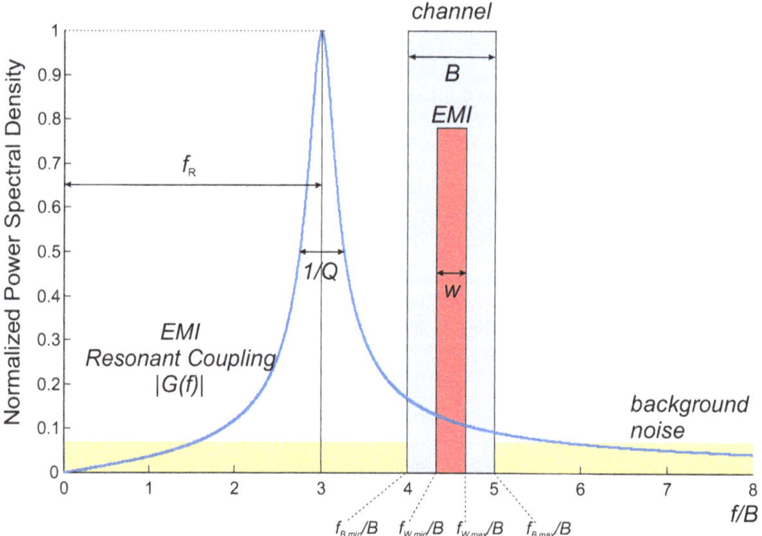

Figure 3. EMI resonant coupling in the frequency domain: power spectral densities of nominal signals, EMI, background noise and resonant coupling transfer function.

In the framework of Shannon information theory, the upper bound of the information than can be reliably transmitted over the AWGN channel considered above in the unit time, is given by the Shannon-Hartley equation [43]:

$$C_0 = B \log_2 \left(1 + \frac{P_S}{BN_0}\right) = B \log_2 (1 + \alpha) \tag{1}$$

where

$$\alpha = \frac{P_S}{BN_0}$$

is the signal-to-noise ratio (SNR) of the AWGN channel.

2.2. Modelling of a Communication Channel Affected by EMI

In the above model, it is now assumed that the communication channel is also corrupted by EMI generated by an aggressor whose operation is based on periodic or randomly SS-modulated switching signals.

In case of periodic signals, the EMI spectrum has nonzero components only at the harmonics kf_0, with $k \in \mathbb{N}$, of the switching frequency f_0 of the aggressor. By contrast, when SS techniques are applied, the instantaneous frequency f of the aggressor varies according to the law

$$f(t) = f_0 + \delta f_0 \xi(t) \tag{2}$$

where f_0 is the central frequency, δ is the modulation depth and $0 \leq \xi(t) \leq 1$ is the modulation profile. As a consequence, the EMI spectral power is more or less uniformly spread over the bandwidth $[kf_0, kf_0(1+\delta)]$ and the spectral peaks are consequently reduced. By using SS-modulations with a random modulation profile $\xi(t)$ optimized for this purpose, the EMI power can be effectively spread over the whole spreading bandwidth $[kf_0, kf_0(1+\delta)]$ in a nearly uniform way, leading to a reduction of the EMI p.s.d. to $P_k/k\delta f_0$ which makes it easier to comply with EMC regulations.

Under the above hypotheses, EMI around an harmonic kf_0 within the bandwidth B of the communication channel in Figure 2 is modelled by a wide-sense stationary (WSS) narrowband Gaussian process, independent both of the background noise and of the transmitted signal, with total power P_{EMI} and bandwidth W, completely or partially overlapping the signal bandwidth B.

2.3. Resonant Coupling Modeling

Even if realistic EMI coupling transfer functions can be rather complex depending on the specific physical mechanism involved in conducted/radiated EMI propagation, focusing the attention to the relatively narrow region of the spectrum included in the bandwidth of the victim channel, practical resonant coupling mechanisms can be conveniently described by a second-order resonant transfer function:

$$G(f) = G_0 \frac{\frac{\jmath f}{Q f_{\text{R}}}}{1 + \frac{\jmath f}{Q f_{\text{R}}} - \frac{f^2}{f_{\text{R}}^2}} \tag{3}$$

where G_0 is a frequency-independent coupling factor, f_{R} is the resonance frequency and Q is the quality factor.

In view of that, for the sake of simplicity and without loss of generality, the effects of resonant EMI coupling on the EMI-induced capacity loss will be discussed in the following considering the EMI transfer function $G(f)$ in (3).

3. EMI-Induced Channel Capacity Loss

Under the hypotheses introduced in the previous Section, the capacity C of a communication channel in the presence of EMI, for a single EMI harmonic overlapping completely or in part with the channel bandwith B, as depicted in Figure 3 can be calculated as

$$C = C_1 + C_2 \tag{4}$$

where C_1 is the capacity of sub-channel where the signal bandwidth overlaps with the EMI bandwidth, i.e., over the bandwidth

$$W^{\star} = B \bigcap W = [W^{\star}_{\text{min}}, W^{\star}_{\text{max}}] = \left[\max\left(f_{B_{\text{min}}}, f_{W_{\text{min}}}\right), \min\left(f_{B_{\text{max}}}, f_{W_{\text{max}}}\right)\right]. \tag{5}$$

and can be expressed as

$$C_1 = \int_{W^\star} \log_2\left(1 + \frac{S_S(f)}{S_N(f) + |G(f)|^2 S_{EMI}(f)}\right) df \qquad (6)$$

where:

$$S_S(f) = \frac{P_S}{B}\Pi_B(f - f_{B,min})$$

is the signal p.s.d.,

$$S_N(f) = N_0$$

is the background noise p.s.d.,

$$S_{EMI}(f) = \frac{P_{EMI}}{W}\Pi_W(f - f_{W,min})$$

is the EMI p.s.d. and $G(f)$ is the resonant coupling transfer function defined in (3).

By introducing in (6) the expressions of $S_S(f)$, $S_N(f)$ and $G(f)$ derived above, the capacity of the sub-channel C_1 affected by EMI can be explicitly evaluated as

$$\begin{aligned} C_1 &= \int_{B\cap W} \log_2\left(1 + \frac{S_S(f)\left[1 + \left(\frac{1}{Q^2} - 2\right)\frac{f^2}{f_R^2} + \frac{f^4}{f_R^4}\right]}{S_N(f) + \left[S_N(f)\left(\frac{1}{Q^2} - 2\right), + \frac{G_0^2}{Q^2}S_{EMI}(f)\right]\frac{f^2}{f_R^2} + S_N(f)\frac{f^4}{f_R^4}}\right) df \\ &= \int_{B\cap W} \log_2\left(1 + \frac{P_S}{BN_0}\frac{1 + \left(\frac{1}{Q^2} - 2\right)\frac{f^2}{f_R^2} + \frac{f^4}{f_R^4}}{1 + \left(\frac{1}{Q^2} - 2 + G_0\frac{P_{EMI}}{WN_0}\right)\frac{f^2}{f_R^2} + \frac{f^4}{f_R^4}}\right) df \\ &= \int_{B\cap W} \log_2\left(1 + \alpha\frac{1 + \left(\frac{1}{Q^2} - 2\right)\frac{f^2}{f_R^2} + \frac{f^4}{f_R^4}}{1 + \left(\frac{1}{Q^2} - 2 + \alpha\beta\right)\frac{f^2}{f_R^2} + \frac{f^4}{f_R^4}}\right) df \\ &= \int_{B\cap W} \log_2(1+\alpha) + \log_2\left(\frac{1 + a_1\frac{f^2}{f_R^2} + \frac{f^4}{f_R^4}}{1 + a_2\frac{f^2}{f_R^2} + \frac{f^4}{f_R^4}}\right) df \end{aligned} \qquad (7)$$

where

$$a_1 = \frac{\frac{1}{Q^2} - 2 + \alpha(\beta + 1)}{1 + \alpha},$$

$$a_2 = \frac{1}{Q^2} - 2 + \alpha\beta$$

and where α is the signal-to-noise ratio as in (1) and $\beta = P_{EMI}/P_S$ is the overall EMI-to-signal power ratio. Since the integral in (7) can be calculated analytically in closed form in terms of:

$$\begin{aligned} \Theta(x,a) &= \int \log_2\left(x^2 + a\,x + 1\right) dx \\ &= \frac{1}{\log 2}\left[\left(x + \frac{a}{2}\right)\log\left(x^2 + ax + 1\right) - 2x + \sqrt{4 - a^2}\arctan\frac{2x - a}{\sqrt{4 - a^2}}\right] \end{aligned} \qquad (8)$$

the capacity C_1 can be finally expressed as

$$C_1 = \frac{W^\star}{B}C_0 + \Theta\left[\left(\frac{W^*_{max}}{f_R}\right)^2, a_1\right] - \Theta\left[\left(\frac{W^*_{min}}{f_R}\right)^2, a_1\right] - \Theta\left[\left(\frac{W^*_{max}}{f_R}\right)^2, a_2\right] + \Theta\left[\left(\frac{W^*_{min}}{f_R}\right)^2, a_2\right] \qquad (9)$$

where C_0 is the capacity of the EMI-free channel.

With the same notations, the capacity C_2 of the complementary sub-channel $B - B \bigcap W$, not affected by EMI, can be expressed as

$$\begin{aligned} C_2 &= \int_{B-B\bigcap W} \log_2 \left(1 + \frac{S_S(f)}{N_0}\right) df \\ &= (B - W^\star) \log_2 (1+\alpha) = \left(1 - \frac{W^\star}{B}\right) C_0. \end{aligned} \tag{10}$$

Replacing (7) and (10) in (4), the overall channel capacity in the presence of EMI can be expressed in the form

$$C = C_0 + \Delta C \tag{11}$$

where

$$\Delta C = \Theta\left[\left(\frac{W^*_{\mathrm{max}}}{f_{\mathrm{R}}}\right)^2, a_1\right] - \Theta\left[\left(\frac{W^*_{\mathrm{min}}}{f_{\mathrm{R}}}\right)^2, a_1\right] - \Theta\left[\left(\frac{W^*_{\mathrm{max}}}{f_{\mathrm{R}}}\right)^2, a_2\right] + \Theta\left[\left(\frac{W^*_{\mathrm{min}}}{f_{\mathrm{R}}}\right)^2, a_2\right] \tag{12}$$

is the channel capacity loss due to the presence of EMI.

The analysis presented so far can be applied to calculate the capacity loss ΔC_k in an AWGN channel due to EMI at frequency kf_0 with up-spreading, symmetric and down-spreading SS-modulations and modulation depth δ, by considering $W = \delta k f_0$ and appropriate values of $W^\star_{\mathrm{min}}$ and $W^\star_{\mathrm{max}}$ in (12) by the same approach detailed in [37]. Moreover, the EMI-induced capacity loss ΔC_{TOT} due to several non-overlapping EMI spectral lines can be immediately evaluated by superposition of the capacity loss contributions due to each spectral line $k = N_1 \dots N_2$ in the channel bandwidth as

$$\Delta C_{\mathrm{TOT}} = \sum_{k=N_1}^{N_2} \Delta C_k. \tag{13}$$

4. Discussion

Based on the results presented in Section 3, the effects of SS-modulation under resonant EMI coupling are now investigated and discussed.

For this purpose, a bandpass channel with bandwidth B ranging from $f_{B_{\mathrm{min}}} = 4B$ to $f_{B_{\mathrm{max}}} = 5B$ is considered and an aggressor generating periodic or random SS-modulated EMI with a fixed total power 20dB below the total signal power, concentrated in a spectral line at frequency f_0 falling in correspondence of the central frequency of the channel $f_C = \frac{f_{B_{\mathrm{min}}} + f_{B_{\mathrm{max}}}}{2} = f_0 = 4.5B$ is considered, as depicted in Figure 4. The signal-to-background noise ratio of the channel is assumed to be 47dB so that the background thermal noise is negligible compared to EMI.

Moreover, it is assumed that the EMI source is coupled to the victim receiver by a resonant coupling function $G(f)$ in (3), with resonant frequency f_{R} and quality factor Q, where the scaling constant G_0 is chosen so that to have unity peak amplitude

$$\max_f |G(f)| = 1.$$

The channel capacity is evaluated accordingly as in Section 3 and is plotted in Figure 4 for $Q = 100$ versus the resonant frequency f_{R} and for different values of the SS EMI bandwidth W normalized with respect to the channel bandwidth B, for W/B ranging from 0 (no SS) up to 25%. Based on (2), depending on the EMI harmonic order k, the SS-modulation depth can be expressed in terms of the normalized bandwidth W/B as

$$\delta = \frac{W}{kf_0} = \frac{W}{B} \frac{1}{4.5 \cdot k}. \tag{14}$$

The same analysis is repeated in Figure 5 for different values of the quality factor Q ranging from 1 to 1,000,000 to discuss the effects of SS-modulations on signal capacity under different resonant coupling quality factors.

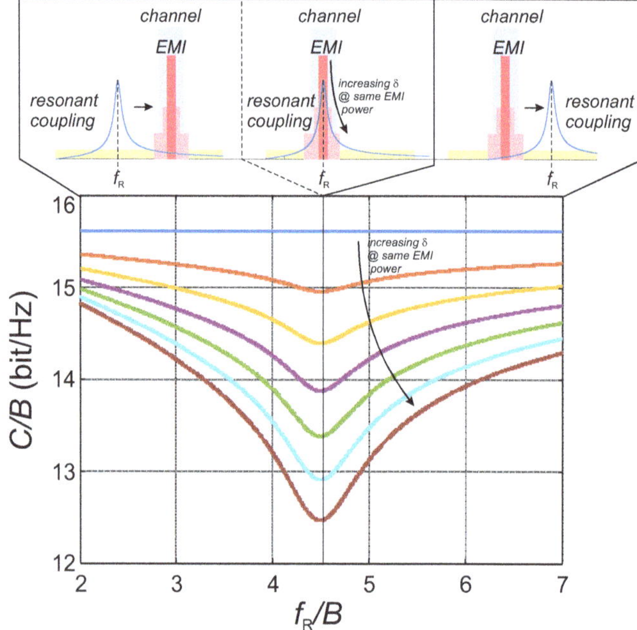

Figure 4. Normalized capacity versus resonant frequency f_R/B for a communication channel with bandwidth $B = (f_1, f_2)$ in the presence of EMI .

For $Q = 1$, the resonant behaviour is weak and the situation is similar to that considered in [37] for wideband EMI coupling (In this study, just the case of EMI bandwidth fully included in the communication channel bandwidth is considered. For extremely narrow-bandwidth communications, EMI coupling can be always considered nearly uniform over the channel bandwidth even in the presence of resonant coupling and the considerations presented in [37] can be directly extended) and an increasing EMI-induced channel capacity loss, weakly dependent on f_R, is observed for an increased SS spreading bandwdith W/B, i.e., for an increased modulation depth δ, which is consistent with [37].

For an increased quality factor Q, the channel capacity loss decreases since the EMI components far from the resonant frequency are partially filtered. Moreover, the capacity loss is more sensitive to the resonant frequency f_R and expectedly reaches a minimum when f_R is close to the EMI central frequency, i.e., for $f_C \simeq f_R$. Even in this case, however, for Q up to 10,000, the EMI-induced capacity loss gets worse and worse by increasing the spreading bandwidth W/B of SS-modulations. A beneficial effect of SS-modulations, which could be possibly expected in view of the reduced SS-EMI coupled power under resonant conditions, as highlighted in Figure 1, is observed only for extremely high Q values exceeding 10,000, where the peak capacity loss decreases for an increasing spreading bandwidth W/B. Even in these cases, for SS-modulations, a capacity impairment is experienced over a wider bandwidth compared to the case with lower W/B or no SS-modulation. Such extremely high values of the quality factor, however, are not realistic for EMI coupling.

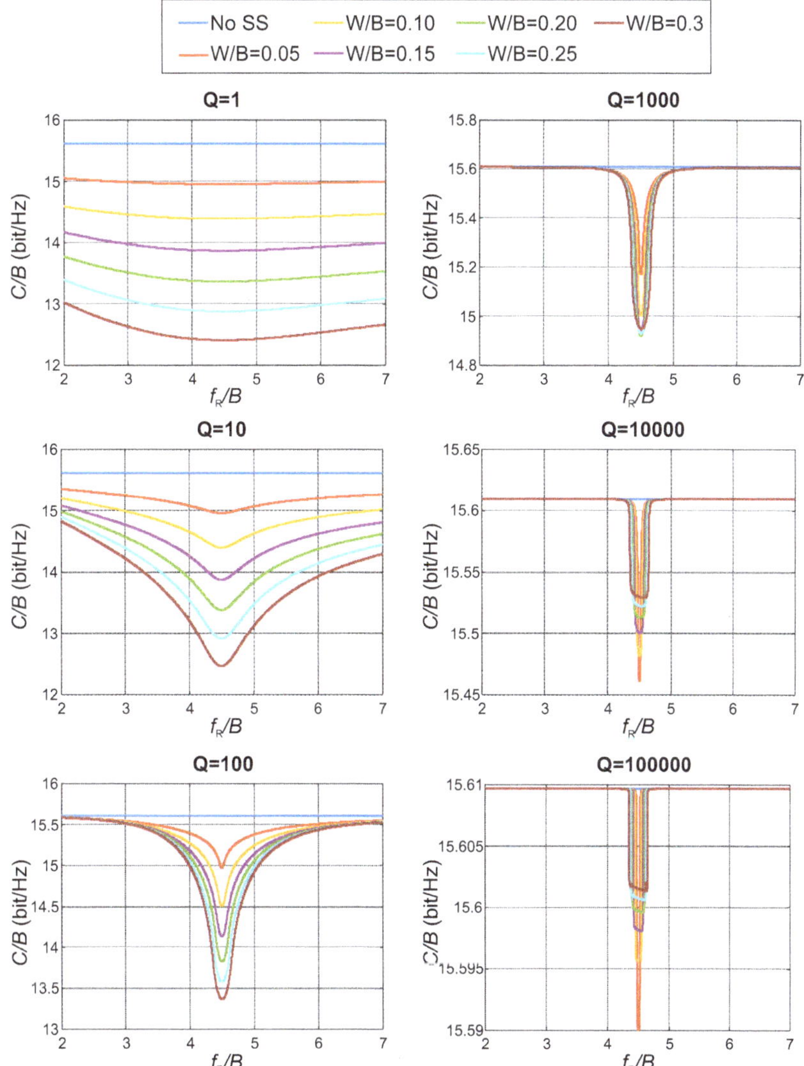

Figure 5. Normalized capacity versus resonant frequency f_R/B for the channel in Figure 3 for different values of the resonant coupling quality factor Q under constant peak coupling, reported in each plot for different SS-modulation depth δ ranging from 0 (no SS) to 30%.

To better highlight the effect of SS-modulations, the worst case channel capacity and the average EMI-induced capacity loss over different values of f_R ranging from $2B$ to $7B$ are reported in Figures 6 and 7, respectively, versus W/B. It can be clearly observed that the worst case capacity monotonically decreases with W/B for practical values of Q below 10,000 and shows an increasing behavior only for extremely high values of Q and for quite large W/B ratios, according to the previous discussion. By contrast, the average channel capacity has found to be monotonically decreasing with W/B for any value of Q.

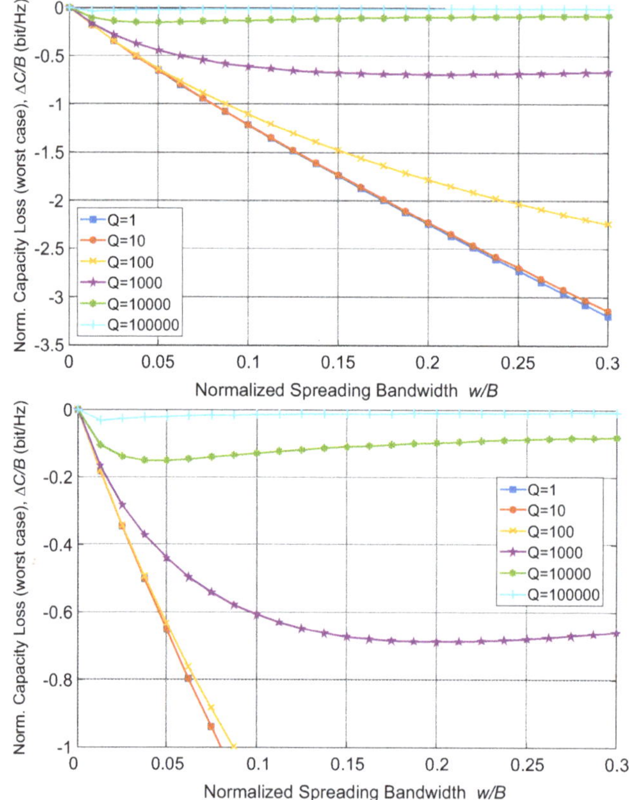

Figure 6. Worst case normalized capacity loss over different resonant frequencies f_R versus SS-modulation depth δ ranging from 0 (no SS) to 30% for different values of the resonant coupling quality factor Q under constant peak coupling. A detail of the top figure is reported in the bottom figure.

To further investigate the effect of SS-modulation under resonant coupling, the analysis described above has been repeated choosing the scaling constant G_0 in (3) so that the coupling function $G(f)$ has unitary energy, i.e., imposing

$$\int_{-\infty}^{+\infty} |G(f)|^2 \mathrm{d}f = 1.$$

In analogy with Figures 6 and 7, the average and the worst case capacity loss over different values of f_R are reported in Figure 8 versus the SS bandwidth W normalized with respect to the channel bandwidth. In this case, a monotonic decrease of both the average and of the worst case channel capacity is observed for increased W/B for all the considered values of the quality factor Q.

Under the hypotheses and limitations considered in the proposed analysis, i.e., that SS EMI can be described as a band-limited Gaussian random process, which is coupled to a wideband AWGN channel by resonant coupling described by the transfer function (3), it can be observed that the application of SS-modulations on periodic interfering signals gives rise to a larger EMI-induced capacity loss under the same EMI total power compared to non-SS-modulated EMI for almost all practical resonant EMI coupling conditions. In other words, the results on the impact of SS-modulations on EMI-induced capacity loss studied in [37] for wideband EMI coupling can be extended to practical resonant coupling conditions according to the proposed analysis.

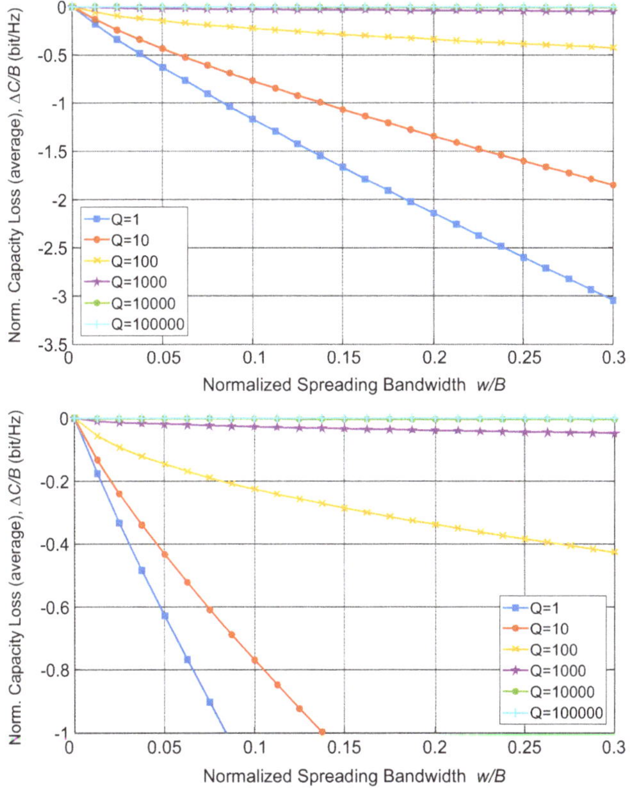

Figure 7. Average normalized capacity for resonant frequencies f_R ranging from $2B$ to $7B$ versus SS-modulation depth δ ranging from 0 (no SS) to 30% for different values of the resonant coupling quality factor Q under constant peak coupling. A detail of the top figure is reported in the bottom figure.

As observed in [37], such a larger capacity loss is not always related to an increased bit error rate (BER) in communication channels operating at sub-capacity bit rates, by the way it is expected to give rise to a worse and worse BER degradation as far as the data rate approaches the Shannon capacity limit, as in real world communication channels employing advanced channel coding schemes (e.g., Turbo coding). Under this perspective, the impact of SS-modulations on the BER of a 4-QAM digital link featuring Turbo coding will be considered as a test case in what follows.

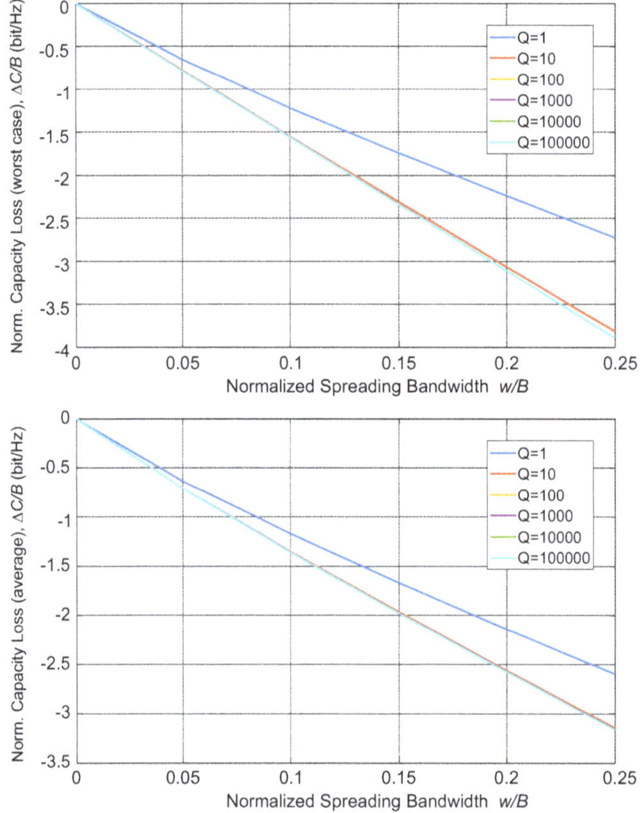

Figure 8. Worst case and average normalized capacity loss over different resonant frequencies f_R versus SS-modulation depth δ ranging from 0 (no SS) to 25% for different values of the resonant coupling quality factor Q under constant coupling integral.

5. Impact of Spread Spectrum Modulations in a 4-QAM Channel under Resonant EMI Coupling

The effects of SS EMI on a communication channel under resonant EMI couping, which have been discussed in the previous section in a completely general, coding-independent way in terms of equivalent channel capacity loss, are now investigated with reference to a synchronous buck DC–DC power converter interfering with a 4-QAM digital link featuring Turbo coding, which makes it possible to achieve a data rate approaching the channel capacity and is therefore expected to be more sensitive to EMI-induced capacity loss [37].

Test Setup

The interference generated by a synchronous buck DC–DC converter has been measured and added in simulation to the input signal of a 4-QAM digital communication channel under resonant coupling with $Q = 1$ and $Q = 100$.

In order to extract the EMI waveforms, the test setup in Figure 9, which is the same presented in [35], is considered. Here, a DC–DC power converter is intentionally designed to interfere with a digital data link. The converter is a hard-switched synchronous Buck operated from an input voltage $V_{IN} = 16$ V and connected to a load $R_L = 10\ \Omega$ in open-loop mode with a fixed duty cycle $D = 0.375$. Such a converter is driven by a 100kHz PWM signal generated by a microcontroller, which can programmed to operate at constant frequency or with an SS random frequency modulation with different modulation depth δ. The EMI voltage v_c at the receiver input without SS-modulation and with random SS-modulation at different modulation depth δ is acquired by a digital scope at 125 MS/s during the DC–DC converter operation and is stored in a database. The power spectral density of the measured waveform for periodic and random SS-modulated ($\delta = 6\%$) is reported in Figure 10.

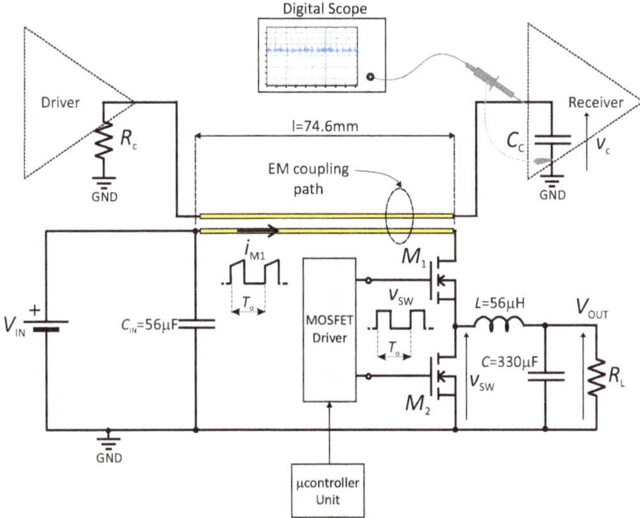

Figure 9. Schematic view of the test setup.

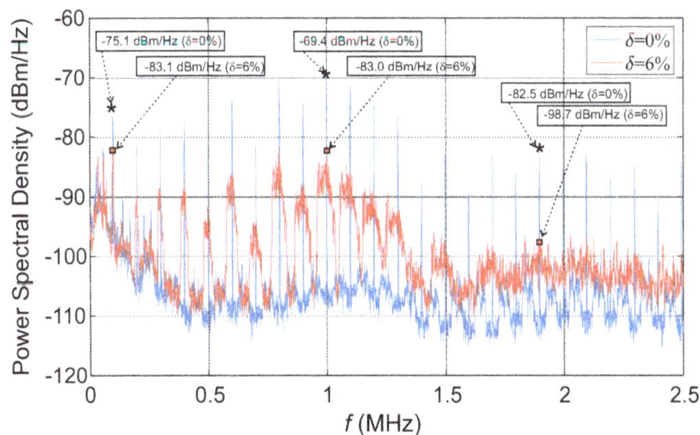

Figure 10. Measured power spectral density of periodic and random SS-modulated ($\delta = 6\%$) EMI.

Then, the measured EMI waveforms are added in simulation to the input of a 4-QAM receiver, modelling the EMI propagation path to the receiver as a resonant transfer function $G(f)$ with $Q = 1$ and $Q = 100$ to discuss the impact of different resonant coupling mechanisms.

For this purpose, a 4-QAM communication scheme over an AWGN channel with $B = 600$ kHz bandwidth is considered and the effects of non-modulated and SS-modulated EMI coupled to the channel bandwidth with a resonant transfer function with $Q = 1$ and with $Q = 100$ have been simulated in the Matlab environment, adding the measured non-modulated and SS-modulated EMI waveforms generated by the power converter to the received input signal.

In Figure 11, resonant coupling with $Q = 1$ is considered and the BER is plotted versus the r.m.s. EMI amplitude (disturbances with different amplitudes have been obtained applying a scaling factor to the same measured EMI waveforms) for different values of the modulation depth δ. From the figure, it can be observed that the EMI amplitude at which BER starts increasing in a significant way is lower for a larger SS-modulation depth and higher in the non-modulated case or for small values of δ.

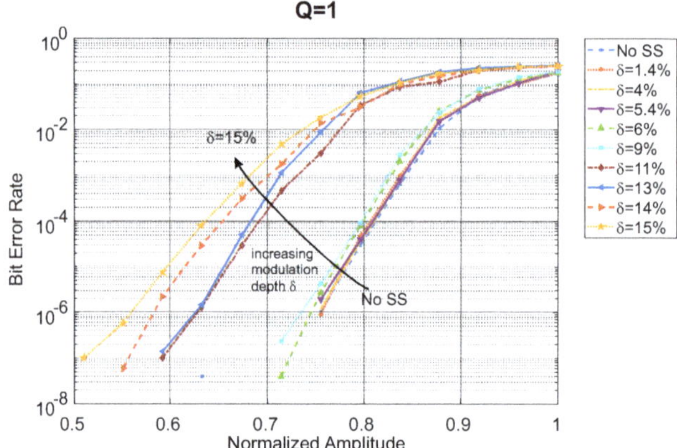

Figure 11. Bit error rate of a communication channel corrupted by SS-modulated EMI under weakly resonant coupling ($Q = 1$) vs. EMI normalized r.m.s. amplitude at different modulation depth δ.

Comparing non-modulated EMI with SS-modulated EMI, a similar BER is achieved for a 20% higher EMI amplitude in the non-modulated case. In Figure 12, the same analysis is performed under resonant EMI coupling with quality factor $Q = 100$. In this case, a similar BER is achieved for a 3.4X higher EMI amplitude than in the non-modulated cases, revealing an even worse impact of SS-modulations on the BER when an EMI resonant coupling with a higher Q factor is introduced.

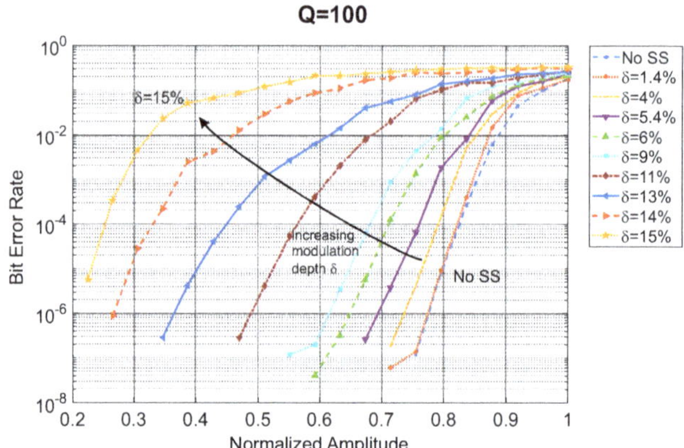

Figure 12. Bit error rate of a communication channel corrupted by SS-modulated EMI under strongly resonant coupling ($Q = 100$) vs. EMI normalized r.m.s. amplitude at different modulation depth δ.

Based on the same analysis described above, the BER has been plotted for different EMI power levels versus the modulation depth δ in Figure 13 for $Q = 1$ and in Figure 14 for $Q = 100$ and also in this case a worse BER degradation with increased SS-modulation depth is observed for resonant coupling with an increased quality factor.

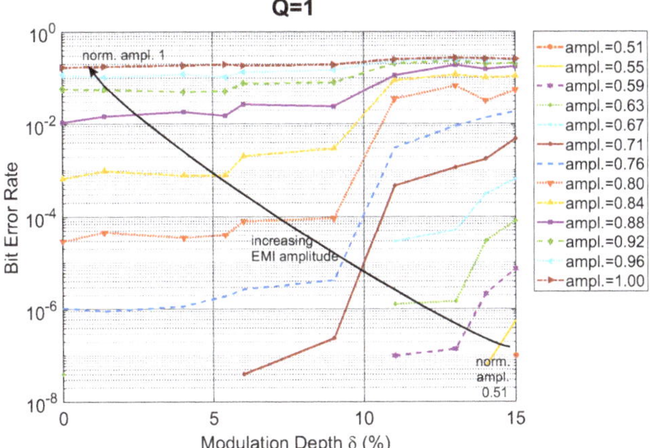

Figure 13. Bit error rate of a communication channel corrupted by SS-modulated EMI under weakly resonant coupling ($Q = 1$) vs. different modulation depth δ at different EMI normalized r.m.s. amplitude values.

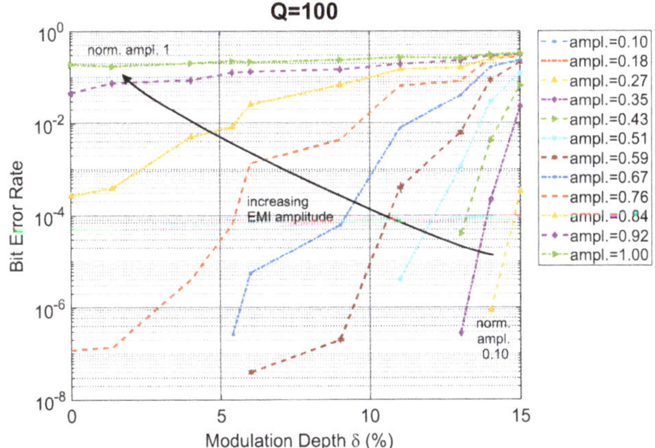

Figure 14. Bit error rate of a communication channel corrupted by SS-modulated EMI under strongly resonant coupling ($Q = 100$) vs. different modulation depth δ at different EMI normalized r.m.s. amplitude values.

6. Conclusions

The effects of EMI coupled to a communication channel by narrowband resonant mechanisms have been investigated in terms of an equivalent channel capacity loss. In this framework, the effects of non-modulated and random SS-modulated EMI under different resonant frequency and quality factors of the EMI coupling mechanism have been compared, thus extending the analysis presented in [37], which was limited to the case of wideband EMI coupling. The analysis has revealed a higher EMI-induced capacity loss for SS-modulated compared to non modulated EMI under practical values of the quality factor Q, while a modest improvement in the worst case capacity loss is observed only for impractical values of Q exceeding 10,000. Simulations on a 4-QAM digital link featuring Turbo coding under EMI resonant coupling have also revealed that SS-modulated EMI gives rise to higher BER at lower EMI power compared WITH non-modulated EMI also in the presence of resonant coupling with practical values, thus confirming a worse interfering potential of SS-modulated EMI.

Author Contributions: Conceptualization, methodology, formal analysis: P.S.C.; validation, writing F.M. All authors have read and agreed to the published version of the manuscript.

Funding: This research received no external funding.

Conflicts of Interest: The authors declare no conflict of interest.

References

1. Dhia, S.B.; Ramdani, M.; Sicard, E. *Electromagnetic Compatibility of Integrated Circuits: Techniques for Low Emission and Susceptibility*; Springer: New York, NY, USA, 2006.
2. Archambeault, B.; Ruehli, A.E. Analysis of power/ground-plane EMI decoupling performance using the partial-element equivalent circuit technique. *IEEE Trans. Electromagn. Compatibil.* **2001**, *43*, 437–445. [CrossRef]
3. Shringarpure, K.; Pan, S.; Kim, J.; Fan, J.; Achkir, B.; Archambeault, B.; Drewniak, J.L. Sensitivity Analysis of a Circuit Model for Power Distribution Network in a Multilayered Printed Circuit Board. *IEEE Trans. Electromagn. Compatibil.* **2017**, *59*, 1993–2001. [CrossRef]
4. Kim, J.; Rotaru, M.D.; Baek, S.; Park, J.; Iyer, M.K.; Kim, J. Analysis of noise coupling from a power distribution network to signal traces in high-speed multilayer printed circuit boards. *IEEE Trans. Electromagn. Compatibil.* **2006**, *48*, 319–330. [CrossRef]
5. Crovetti, P.S. Operational amplifier immune to EMI with no baseband performance degradation. *Electron. Lett.* **2010**, *46*, 209–210. [CrossRef]
6. Redouté, J.; Richelli, A. A methodological approach to EMI resistant analog integrated circuit design. *IEEE Electr. Comp. Mag.* **2015**, *4*, 92–100. [CrossRef]
7. Villavicencio, Y.; Musolino, F.; Fiori, F. Electrical Model of Microcontrollers for the Prediction of Electromagnetic Emissions. *IEEE Trans. Very Large Scale Integr. (VLSI) Syst.* **2011**, *19*, 1205–1217. [CrossRef]
8. 'Code of Federal Regulations, Title 47, Chapter I, Subchapter A, Part 15, Subpart B: Unintentional Radiators'. Available online: https://www.ecfr.gov/ (accessed on 19 July 2018).
9. *Council Directive 89/336/EEC on the Approximation of the Laws of the Member States Relating to Electromagnetic Compatibility*; Official Journal of the European Union, Technical Report L139/19; The Publications Office of the European Union: Luxembourg, 1989.
10. *CISPR 11:2015 Industrial, Scientific and Medical Equipment - Radio-Frequency Disturbance Characteristics—Limits and Methods of Measurement*; IEC: Geneva, Switzerland, 2015.
11. *CISPR 14-1:2016 Electromagnetic Compatibility - Requirements for Household Appliances, Electric Tools and Similar Apparatus—Part 1: Emission*; IEC: Geneva, Switzerland, 2016.
12. Musolino, F.; Villavicencio, Y.; Fiori, F. Chip-Level Design Constraints to Comply With Conducted Electromagnetic Emission Specifications. *IEEE Trans. Electromagn. Compatibil.* **2012**, *54*, 1137–1146. [CrossRef]
13. Crovetti, P.S.; Fiori, F.L. Efficient BEM-based substrate network extraction in silicon SoCs. *Microelectron. J.* **2008**, *39*, 1774–1784. [CrossRef]
14. Lobsiger, Y.; Kolar, J.W. Closed-Loop di/dt and dv/dt IGBT Gate Driver. *IEEE Trans. Power Electron.* **2015**, *30*, 3402–3417. [CrossRef]
15. Chung, H.; Hui, S.Y.R.; Tse, K.K. Reduction of power converter EMI emission using soft-switching technique. *IEEE Trans. Electromagn. Compatibil.* **1998**, *40*, 282–287. [CrossRef]
16. Pareschi, F.; Rovatti, R.; Setti, G. EMI Reduction via Spread Spectrum in DC/DC Converters: State of the Art, Optimization, and Tradeoffs. *IEEE Access* **2015**, *3*, 2857–2874. [CrossRef]
17. Lin, F.; Chen, D.Y. Reduction of power supply EMI emission by switching frequency modulation. *IEEE Trans. Power Electron.* **1994**, *9*, 132–137.
18. Hardin, K.B.; Fessler, J.T.; Bush, D.R. Spread spectrum clock generation for the reduction of radiated emissions. In Proceedings of the IEEE Symposium on Electromagnetic Compatibility, Chicago, IL, USA, 22–26 August 1994; pp. 227–231.
19. Chen, Y.; Ma, D.B. 15.7 An 8.3MHz GaN Power Converter Using Markov Continuous RSSM for 35dBμV Conducted EMI Attenuation and One-Cycle TON Rebalancing for 27.6dB VO Jittering Suppression. In Proceedings of the 2019 IEEE International Solid- State Circuits Conference—(ISSCC), San Francisco, CA, USA, 17–21 February 2019; pp. 250–252.

20. De Martino, M.; De Caro, D.; Esposito, D.; Napoli, E.; Petra, N.; Strollo, A.G.M. A Standard-Cell-Based All-Digital PWM Modulator With High Resolution and Spread- Spectrum Capability. *IEEE Trans. Circuits Syst. I Regul. Pap.* **2018**, *65*, 3885–3896. [CrossRef]
21. Jun, J.; Bae, S.; Lee, Y.; Kim, C. A Spread Spectrum Clock Generator With Nested Modulation Profile for a High-Resolution Display System. *IEEE Trans. Circuits Syst. II Express Briefs* **2018**, *65*, 1509–1513. [CrossRef]
22. De Caro, D.; Tessitore, F.; Vai, G.; Imperato, N.; Petra, N.; Napoli, E.; Parrella, C.; Strollo, A.G. A 3.3 GHz Spread-Spectrum Clock Generator Supporting Discontinuous Frequency Modulations in 28 nm CMOS. *IEEE J. Solid-State Circuits* **2015**, *50*, 2074–2089. [CrossRef]
23. Hwang, S.; Song, M.; Kwak, Y.; Jung, I.; Kim, C. A 3.5 GHz Spread-Spectrum Clock Generator With a Memoryless Newton-Raphson Modulation Profile. *IEEE J. Solid-State Circuits* **2012**, *47*, 1199–1208. [CrossRef]
24. *Spread Spectrum Clocking Using the CDCS502/503*; Application Note SCAA103; Texas Instruments: Dallas, TX, USA, 2009.
25. *STM32 MCUs Spread-Spectrum Clock Generation Principles, Properties and Implementation*; Application Note AN4850; STMicroelectronics: Geneva, Switzerland, 2016.
26. *R1275S Series, 30V, 2A, Synchronous PWM Step-Down DC/DC Converter*; Ricoh Electronic Devices Co. Ltd.: Tokyo, Japan, 2018.
27. Multiphase Oscillator with Spread Spectrum Frequency Modulation, Document LTC6902, Linear Technology, 2003. Available online: http://cds.linear.com/docs/en/datasheet/6902f.pdf (accessed on 19 December 2019).
28. Callegari, S.; Rovatti, R.; Setti, G. Spectral properties of chaos-based FM signals: theory and simulation results. *IEEE Trans. Circuits Syst. I Fundament. Theory Appl.* **2003**, *50*, 3–15. [CrossRef]
29. Setti, G.; Balestra, M.; Rovatti, R. Experimental verification of enhanced electromagnetic compatibility in chaotic FM clock signals. In Proceedings of the 2000 IEEE International Symposium on Circuits and Systems (ISCAS), Geneva, CH, 28–31 May 2000; Volume 3, pp. 229–232.
30. KTse, K.; Chung, H.S.; Hui, S.Y.R.; So, H.C. A comparative study of carrier-frequency modulation techniques for conducted EMI suppression in PWM converters. *IEEE Trans. Ind. Electron.* **2002**, *49*, 618–627.
31. Lauder, D.; Moritz, J. *Investigation into Possible Effects Resulting From Dithered Clock Oscillators on EMC Measurements and Interference to Radio Transmission Systems*; Radiocommunications Agency: London, UK, March 2000.
32. Hardin, K.; Oglesbee, R.A.; Fisher, F. Investigation into the interference potential of spread-spectrum clock generation to broadband digital communications. *IEEE Trans. Electromagn. Compatibil.* **2003**, *45*, 10–21. [CrossRef]
33. Mukherjee, R.; Patra, A.; Banerjee, S. Impact of a Frequency Modulated Pulsewidth Modulation (PWM) Switching Converter on the Input Power System Quality. *IEEE Trans. Power Electron.* **2010**, *25*, 1450–1459. [CrossRef]
34. Skinner, H.; Slattery, K. Why spread spectrum clocking of computing devices is not cheating. In Proceedings of the 2001 IEEE EMC International Symposium. Symposium Record. International Symposium on Electromagnetic Compatibility (Cat. No.01CH37161), Montreal, QC, Canada, 13–17 August 2001.
35. Musolino, F.; Crovetti, P.S. Interference of Spread-Spectrum Switching-Mode Power Converters and Low-Frequency Digital Lines. In Proceedings of the 2018 IEEE International Symposium on Circuits and Systems (ISCAS), Florence, Italy, 27–30 May 2018; pp. 1–5.
36. Matsumoto, Y.; Shimizu, T.; Murakami, T.; Fujii, K.; Sugiura, A. Impact of Frequency-Modulated Harmonic Noises From PCs on OFDM-Based WLAN Systems. *IEEE Trans. Electromagn. Compatibil.* **2007**, *49*, 455–462. [CrossRef]
37. Musolino, F.; Crovetti, P.S. Interference of Spread-Spectrum Modulated Disturbances on Digital Communication Channels. *IEEE Access* **2019**, *7*, 158969–158980. [CrossRef]
38. Shim, H.; Hubing, T.H. A closed-form expression for estimating radiated emissions from the power planes in a populated printed circuit board. *IEEE Trans. Electromagn. Compatibil.* **2006**, *48*, 74–81. [CrossRef]
39. Zeeff, T.M.; Hubing, T.H. Reducing power bus impedance at resonance with lossy components. *IEEE Trans. Adv. Pack.* **2002**, *25*, 307–310. [CrossRef]
40. Grassi, F.; Spadacini, G.; Marliani, F.; Pignari, S.A. Use of Double Bulk Current Injection for Susceptibility Testing of Avionics. *IEEE Trans. Electromagn. Compatibil.* **2008**, *50*, 524–535. [CrossRef]

41. Crovetti, P.S.; Fiori, F. Distributed Conversion of Common-Mode Into Differential-Mode Interference. *IEEE Trans. Microw. Theory Tech.* **2011**, *59*, 2140–2150. [CrossRef]
42. Crovetti, P.S. Finite Common-Mode Rejection in Fully Differential Nonlinear Circuits. *IEEE Trans. Circuits Syst. II Express Briefs* **2011**, *58*, 507–511. [CrossRef]
43. Shannon, C.E. A Mathematical Theory of Communication. *Bell Syst. Tech. J.* **1948**, *27*, 379–423. [CrossRef]

Article

Signal Transformations for Analysis of Supraharmonic EMI Caused by Switched-Mode Power Supplies

Leonardo Sandrolini [1,*] and Andrea Mariscotti [2]

1 Department of Electrical, Electronic and Information Engineering (DEI), University of Bologna, 40126 Bologna, Italy
2 Department of Electrical, Electronics and Telecommunication Engineering and Naval Architecture (DITEN), University of Genova, 16145 Genova, Italy; andrea.mariscotti@unige.it
* Correspondence: leonardo.sandrolini@unibo.it; Tel.: +39-051-2093-484

Received: 31 October 2020; Accepted: 4 December 2020; Published: 7 December 2020

Abstract: Switched-Mode Power Supplies (SMPSs) are a relevant source of conducted emissions, in particular in the frequency interval of supraharmonics, between 2 kHz and 150 kHz. When using sampled data for assessment of compliance, methods other than Fourier analysis should be considered for better frequency resolution, compact signal energy decomposition and a shorter time support. This work investigates the application of the Wavelet Packet Transform (WPT) and the Empirical Mode Decomposition (EMD) to measured recordings of SMPS conducted emissions, featuring steep impulses and damped oscillations. The comparison shows a general accuracy of the amplitude estimate within the variability of data themselves, with very good performance of WPT in tracking on stationary components in the low frequency range at some kHz. WPT performance however may vary appreciably depending on the selected mother wavelet for which some examples are given. EMD and its Ensemble EMD implementation show a fairly good accuracy in representing the original signal with a very limited number of base functions with the capability of exploiting a filtering effect on the low-frequency components of the signal.

Keywords: electromagnetic compatibility; conducted emissions; Discrete Wavelet Transform; electromagnetic interference; Empirical Mode Decomposition; harmonics; Switched-Mode Power Supplies; transients; Wavelet Packet Transform

---

1. Introduction

The widespread use of Switched-Mode Power Supplies (SMPSs) can be justified with flexibility, portability and better efficiency, including superior voltage regulation at the dc load. SMPSs feature various types of power electronic converters topologies, with conducted emissions that need to be assessed over a wide frequency range [1]. Conducted emissions caused by the switching process, in fact, originate from switching fundamentals of some tens of kHz and spread in the hundreds kHz and MHz ranges, well above the "classical" interval of harmonic and interharmonic penetration studies, limited in general to some kHz for 50–60 Hz systems. In the last ten years, attention has been extended to frequency intervals below 150 kHz [2–9], in general not currently regulated by electromagnetic compatibility (EMC) standards, focusing on the so-called "supraharmonics". Some standards [10–14] deal with the 2–150 kHz interval, but do not cover explicitly SMPSs and other similar switching converters, thus taking into account the characteristics of the related signals. The approach of EMC standards to the quantification of conducted emissions is quite similar above and below 150 kHz: resolution bandwidth of 9 kHz and frequency sweep by a spectrum analyzer or EMI receiver is the reference approach. However, the use of time-domain sampling is a valid alternative for the many data acquisition boards and digital sampling oscilloscopes available, for the possibility of inspecting the "real" waveform when cross-checking the final results (especially in case of noncompliance) and the possibility of changing and tuning the post-processing methods.

The most popular of these methods is the spectrum calculation by Fourier transform (a Discrete Fourier Transform (DFT) or a Short-Time Fourier Transform (STFT), tracking Fourier spectrum versus time): it is a well-known technique that can be mastered after some practice, but may in reality lead to some disappointment, especially in the case of transient signals with a short time support [15], as it briefly pointed out in Section 2. Alternative methods better suited to handle signals characterized by transients may be considered: wavelets with the Wavelet Packet Transform (WPT) [16–22] and the Empirical Mode Decomposition (EMD), in the Ensemble EMD (EEMD) implementation [23–26] (to avoid the EMD shortcomings, as shown in Section 2). Although they have both been extensively described in the literature and the former applied often to Power Quality analysis, the considered problem is peculiar in two aspects:

- The investigated methods must provide accurate results in terms of amplitude of the signal components that may be compared to existing or future emission limits, in order to assess compliance and related margins. Several wavelet applications do not discuss the amplitude accuracy in the presence of both narrowband and broadband components, as well as of significant variability of the instantaneous frequency.
- They must support the analysis of the signal and of the control measures necessary in the case of noncompliance, with clear relationship with the internal sources and switching mechanisms, as well as also being easily interpretable, at least for the most relevant components critical for compliance to limits.

Modern controllers for switching converters adopt smart modulation techniques to improve efficiency, reduce voltage stress on components and reduce EMI. They are able to rapidly switch between different modulation patterns, each optimized for a range of operations and possibly with different spectral signatures: this opens the door to potential non-stationarity of the signal and the need to adopt short time windows, unsuitable for a Fourier-based approach.

Traditional EMC techniques based on frequency sweep have fallen behind the technological development in some sectors, such as power transfer and conversion [27], as well as in the presence of significant transients as the first source of emissions [28]. They would benefit from time–frequency analysis methods that are computationally attractive and preserve the accuracy necessary for the assessment of compliance. The focus is on amplitude accuracy, with the frequency estimate used only to locate suitable limit values and track component behavior. The definition of a reference case for comparison is particularly challenging, as discussed in Section 3.

This work describes the setup and measurement methods in Section 2. Then, after a synthetic description in Section 3 of the WPT and EMD transforms (with a comparison between EMD and EEMD), results are compared and discussed in Section 4.

2. Measurement Setup and Signal Transformations

2.1. Measurement Setup

The signals considered in this work are the conducted emissions of SMPSs connected to the AC mains through a Line Impedance Stabilization Network (LISN). The output of the LISN is fed to an 8-bit DSO with a sampling rate of 10 MSa/s, to keep records to a manageable size for later post-processing. Frequency domain sweeps can be carried out as well by means of an EMI receiver, to use as reference in case of need. The scheme of the setup is shown in Figure 1.

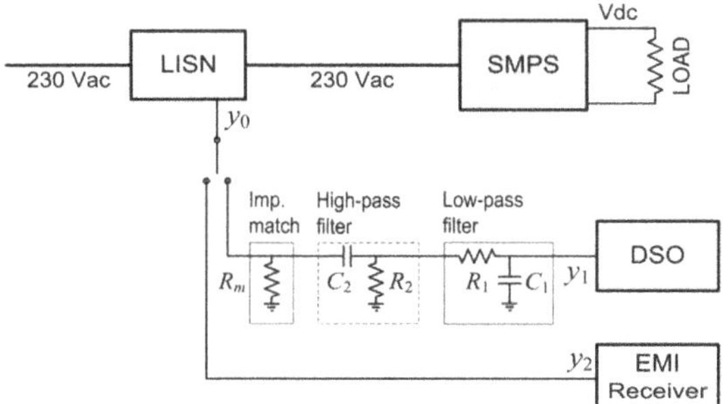

Figure 1. Scheme of the experimental setup: SMPS under test and its dc variable load, LISN (Line Impedance Stabilization Network), 50 Ω impedance matching, high-pass filter (1 kHz cutoff) and low-pass filter (2 MHz cutoff), Digital Storage Oscilloscope (DSO) and EMI Receiver.

An anti-aliasing low-pass filter with a cutoff frequency of 2 MHz is used to feed the DSO: the resistor R_2 is 5 kΩ and the capacitance C_2 is the internal input capacitance of the DSO, equal to 15 pF with good accuracy. A high-pass filter with a cutoff frequency of 1 kHz is also included for additional attenuation of the fundamental and the 100 Hz ripple component possibly leaking through the LISN: the resistor R_1 = 1.5 kΩ is large enough to be neglected in parallel to the impedance-matching 50 Ω resistor R_m, giving a neat 100 nF for C_1 to get the desired 1 kHz cutoff.

The tested SMPSs are all AC/DC low-voltage switching converters with a 5 V or 12 V nominal dc output and a power level in the range of some tens of W. This allowed the use of a 16 A LISN and no major issues of heating and high-voltage hazards; it is not likewise a limiting factor, since it is expected that smaller converters will show the most rapid switching transients. As a matter of fact, the observed switching frequency values are in the range of some tens of kHz, in line with the dynamics of controllers and MOS transistors available about ten years ago.

The switching byproducts responsible for the conducted emissions are superimposed to the mains sinusoidal voltage, located at its zero crossings. The signal y_1 at the DSO input (qualitatively similar to the signal y_0 as provided at the LISN output) is an intermittent waveform, deprived of the 50 Hz mains voltage component and of its low-order harmonics, also to improve the dynamic range. The zero-mean high-pass filtered signal is characterized by steep transients and rapid oscillations, as shown in Figure 2. The instantaneous frequency of oscillations is in the order of 20–30 kHz and is in relation to the internal switching frequency; the repetition rate is 100 Hz, twice the mains frequency.

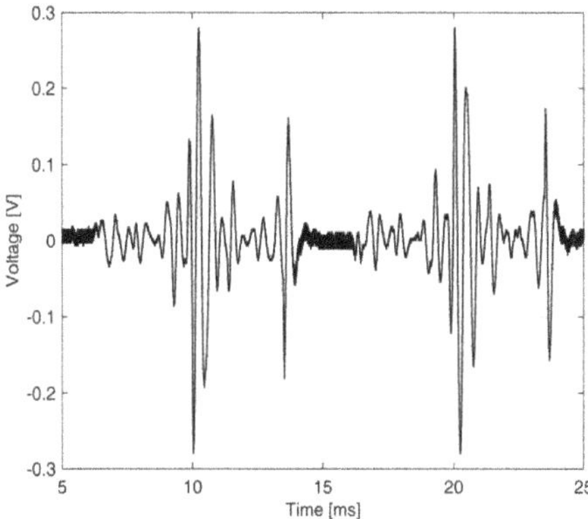

Figure 2. Typical time-domain waveform of an SMPS at the high-pass filtered LISN output measured as DSO input signal y_1.

2.2. Use of the Discrete Fourier Transform (DFT)

The Fourier-based analysis, for simplicity gathered under the abbreviation DFT, is assumed well known and is not repeated. As seen, the signal is characterized by bursts with oscillations at low and high frequency of limited duration. The use of DFT applied to this kind of intermittent rapidly varying signals has the following drawbacks:

- The signal portion to analyze has a short duration, which contrasts to the desirable or required frequency resolution. The IEC 61000-4-7 standard [29] indicates an observation time of 200 ms that is clearly inadequate (equivalent to 5 Hz resolution, suitable for low frequency harmonics, but not for phenomena at higher frequency with fast dynamics, as in the present case). The EN 55065-1 [14] and in general EMC standards for conducted emissions below 150 kHz require a 200 Hz resolution bandwidth. A more suitable approach has been proposed in the IEC 61000-4-30 [13], defining a frequency resolution of 2 kHz, which goes in the direction of tracking signals with fast dynamics, as in the present case.
- To avoid spectral leakage, the signal should be cut in the zero-valued short intervals between pulses, implying a resolution frequency of about 100 Hz. Other window intervals will suffer from spectral leakage, only partly attenuated by the use of tapering windows.
- Using a long window interval has the drawback of averaging the contribution of the contained signal components, largely reducing the estimated amplitude of the peak located at the center.
- Using a short window interval reduces the frequency resolution and worsens the spectral representation and the estimate of the amplitude, this time caused by the short-range spectral leakage (or "picket fence effect"), if no additional post processing is used.

For all the reasons above and as justified in the Introduction, alternative signal transformations should be investigated for applicability to the present case, providing spectral representations in a domain (or combination of domains, e.g., time and frequency) that can still be interpreted and used to evaluate the degree of compliance, as for conducted emissions in the aforesaid 2–150 kHz interval. Wavelets have been extensively applied for accurate PQ analysis in the harmonic low-frequency range [1,16,18–21], as effective alternative method for compliance to the IEC 61000-4-30 processing requirement and for detection of transient PQ phenomena (e.g., voltage sags, overvoltages, etc.) [17], the latter without the analysis of the accuracy of the parameters estimate. EMD and EEMD find their application in the analysis of seismic data, vibrations, speech signals and medical data [23–25], but lack a clear quantification of component amplitude to support comparison with limits and susceptibility criteria, defined in the frequency domain. Rather their typical application is more oriented to the detection and identification of signal features and components hidden by noise and signal compression.

2.3. Wavelet Packet Transform (WPT)

Oppositely to the DFT with its stationary sinusoidal kernel, wavelet analysis relies on short-duration oscillating waveforms that have zero mean and decay rapidly to zero at both ends [16]. Dilation and shifting of kernel waveforms allow adapting it for variable time and frequency resolution. Since its first inception, wavelets have been proposed in both wavelet function and multistage filter bank implementations. The filter bank was built using low-pass (LP) and high-pass (HP) complementary filters (for reference, they can be created as "quadrature mirror filters") [16,18]. The first implementation (Discrete Wavelet Transform (DWT)) applied recursively the LP and HP filters of the successive stage only to the LP output of the previous stage, and so on. The Wavelet Packet Transform (WPT) uses then a symmetrical structure, applying the decomposition to both LP and HP outputs, as shown in Figure 3, which allows a linear, rather than logarithmic, apportionment of the frequency axis, and thus a linear decomposition of the signal spectrum. This is inline with the 2 kHz group representation uniformly spread over the frequency axis suggested by the EN 61000-4-30 [13]. At each stage, the outputs of the LP and HP filters are named "approximation" and "detail", respectively.

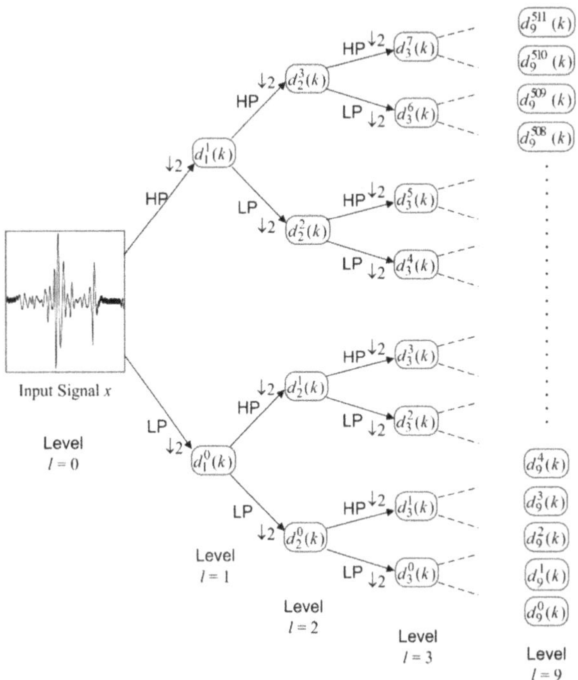

Figure 3. WPT (Wavelet Packet Transform) tree structure, indicating levels, ordinal indices at each level and terminal nodes. The down arrow "↓n" indicates a down sampling (or decimation) by a factor of n (in our case, n = 2).

The output of the WPT decomposition is used for a compact spectrum estimate focused on the determination of three elements: the amplitude of the component, its location on the frequency axis and its location on the time axis (or duration). The STFT addresses this third point by assigning a window length and the amount of overlap between successive windows sliding over the signal along the time axis.

The WPT spectrum estimate is built on the "details" of the terminal nodes, which are in a number $N = 2^L$, where L is the number of chosen levels. Details are denoted as $d_l^p(k)$, where l indicates the level ($l = 0, 1, \dots L$, level 0 being the original signal x), p is the position along the row of nodes in the tree at the same level ($p = 0, 1, \dots 2^l$) and k is a "time" index (the position along the sequence of data samples). Each detail in fact apportions the original signal of length M over the N nodes at a given level with data sequences of length M/N; the lower in the tree, the higher the level l and the number of nodes ($N_l = 2^l$), and the shorter the length of each sequence $K = M/2^l$. Each data sequence contained by a detail (for ease of understanding, we consider the last and deeper level, i.e., the terminal nodes) is a sub-band representation of the original signal: the bandwidth, or frequency resolution, Δf is derived from the original sampling f_s as $\Delta f = f_s/2^{L+1}$, and in general $\Delta f_l = f_s/2^{l+1}$. In our case with a downsampled version of the raw data using $f_s = 1$ MSa/s and $L = 9$, we obtain $\Delta f = 976$ Hz, which matches the selected RBW for the EMI receiver data used for comparison in Section 3.

To the aim of the quantitative assessment of the accuracy of WPT spectrum estimate, the amplitude of the components of the spectral representation must be derived: the Matlab function wpspectrum( ) gives such time–frequency representation, where time and frequency resolutions Δt and Δf are determined by the selected number of levels L and the original sampling f_s.

The problem is to understand the unity of measure and the meaning of the spectrum values (pixels, or tiles) in terms of the original quantities; in other words, WPT spectrum extraction needs to be "calibrated" in order to be used as an EMI assessment tool.

Given the details as sub-banded data sequences, the calculation of the total rms value gives straightforwardly a measure of the power contained in each sub-band p (the notation $l = L$ in (1) indicates that the expression is general for whatever l, but we select in the following $l = L$, so that the terminal nodes give the desired frequency resolution).

$$\widetilde{x}_{p,l=L} = \sqrt{\frac{\sum_k \left(d^p_{l=L}(k)\right)^2}{N_{l=L}}} \tag{1}$$

Of course, rms values from adjacent sub-sequences may be aggregated with a concept of overall rms in a wider band, made of the composition of each respective Δf^p_l (the sub-band of order p at the level l). Being rms values interpreted in a power perspective, such aggregation is performed by rms summation over the sub-sequences $d^p_L(k)$ with p in the selected α set (written as p_α).

$$\widetilde{x}_{[p_\alpha],L} = \sqrt{\frac{\sum_\alpha \sum_k \left(d^{p_\alpha}_L(k)\right)^2}{N_L}} \tag{2}$$

To exemplify, to make an approximately 2 kHz representation from the $\Delta f = 976$ Hz representation, two adjacent sub-bands must be aggregated, discarding the first one that contains the dc component. Thus, the p indices would start from 1 and go in pairs: (2,3), (4,5), (6,7), etc. A bandwidth of approximately 3 kHz is obtained by grouping with a ratio of 3, so (2,3,4), (5,6,7), etc.

2.4. Empirical Mode Decomposition (EMD) and Ensemble Empirical Mode Decomposition (EEMD)

The Empirical Mode Decomposition (EMD) was proposed for attractive characteristics: the transformation is based on simple operations (and should be relatively light computationally), its results can be clearly interpreted and it is robust, thus there are no wide or uncontrolled variations of the results for small changes of the parameters.

The EMD was developed to prepare the data to which the Hilbert Transform (HT) could be applied; the combination of EMD and HT is the Hilbert-Huang Transform (HHT), which provides a powerful tool for nonlinear and nonstationary signal processing. The signal is decomposed adaptively into a finite (often small) number of the so-called Intrinsic Mode Functions (IMFs), not known a priori (differently than in the Fourier or Wavelet analysis), which represent the oscillation modes embedded in the data. With the HT, the IMFs yield instantaneous frequencies as functions of time. Each IMF satisfies the following two conditions:

- In the whole dataset, the number of extrema and the number of zero crossings is either equal or differ at least by one.
- At any point, the mean value of the envelope defined by local maxima and local minima is zero.

The iterative process of extraction of the IMFs through the EMD method is called the sifting process. The first IMF contains the highest oscillation frequency in the signal. The difference between the original signal and the first IMF is called a residue. The residue is then considered as new data to decompose and the sifting process is applied to it; the same definition thus applies to the difference between the previous residue and the last IMF at later stages. The sifting process must be repeated until the extracted signal (the candidate IMF) satisfies the IMF definition or the predefined maximum number of iterations is exceeded. The number of extrema decreases as the decomposition proceeds, and the sifting process ends when the residue becomes a monotonic function or a function with only one extremum from which no further IMF can be extracted [23] (in other words, no oscillations are contained in the residue). At the end of the decomposition, we have N IMF functions and one residue, and the original signal $x(t)$ can be reconstructed as

$$x(t) = \sum_{i=1}^{N} c_i(t) + r(t) \tag{3}$$

where c_1 and c_N are the highest and lowest frequency IMFs, respectively, and $r(t)$ is the residue. Besides the harmonic content, IMFs are characterized by different amplitudes. The original signal may then be fairly approximated by using only a limited set of the IMFs, i.e., those with the largest amplitudes only.

The major drawback of the EMD method is mode mixing, for which the EEMD method was introduced [26]. Mode mixing means that different modes of oscillations coexist in the same IMF. EEMD (Ensemble EMD) consists in adding white noise to the original signal. This procedure is run for a number of times, thus an ensemble of datasets with different added white noise is generated. The new obtained data are then decomposed into IMFs with the sifting procedure. Being the added noise different in each run, the IMFs of the various runs are uncorrelated. Averaging the IMFs eliminates the added noise and yields then the final result. The truth IMF, $c_j(t)$, defined by EEMD is then obtained for an ensemble number approaching infinity:

$$c_j(t) = \lim_{N\to\infty} \sum_{k=1}^{N} c_{jk}(t) \tag{4}$$

where

$$c_{jk}(t) = c_j(t) + r_{jk}(t) \tag{5}$$

is the kth trial of the jth IMF in the noise added signal, being $r_{jk}(t)$ the contribution from the added white noise of the kth trial to the jth IMF. To minimize the difference between the truth IMF $c_j(t)$ and the IMF $c_j(t)$ obtained with N trials, N must be large, as for the well-known statistical rule the difference between the truth and the result of the ensemble for a finite number of elements of the ensemble decreases as $1/\sqrt{N}$.

EEMD represents a major improvement to EMD as it eliminates mode mixing; however, the major drawback is that the result of EEMD does not strictly satisfy the definition of IMF, as the sum of IMFs is not necessarily an IMF. A possible solution to this problem is to apply EMD to the IMFs produced by EEMD.

3. Results

3.1. Introduction and Reference Case

Regarding the accuracy of the estimated amplitude, it is difficult to establish a reference case for this kind of signals. In other words, what is the real amplitude of the spectrum components of the signal?

The STFT gives a first indication, although the amplitude is highly variable, as a function of the frequency resolution, its implicit averaging and the amount of spectral leakage depending on the characteristics of the adopted tapering window.

It may be objected that an EMI receiver scan returns the reference spectrum, since after all it is the reference method adopted by EMC standards, against which we check the compliance of emissions. The "EMC" resolution bandwidth (RBW) values of 200 Hz and 9 kHz used, respectively, below and above 150 kHz have been considered as a starting point, assessing the variability of the measured spectrum by changing RBW from 200 Hz to 1 kHz, 2 kHz, 3 kHz and 5 kHz, looking for a compromise between a neat frequency resolution and an accurate amplitude estimate. The results are reported in Figure 4 showing a progressive increase of the noise floor (as expected), a substantially stable amplitude of the switching peaks at 43.3 kHz and 87 kHz and above, and a variable estimated amplitude for the low-frequency nonstationary components (with a behavior similar to that of background noise). The result with 1 kHz RBW is retained for further analysis as a good compromise, also in view of the necessary time resolution settings in STFT and WPT.

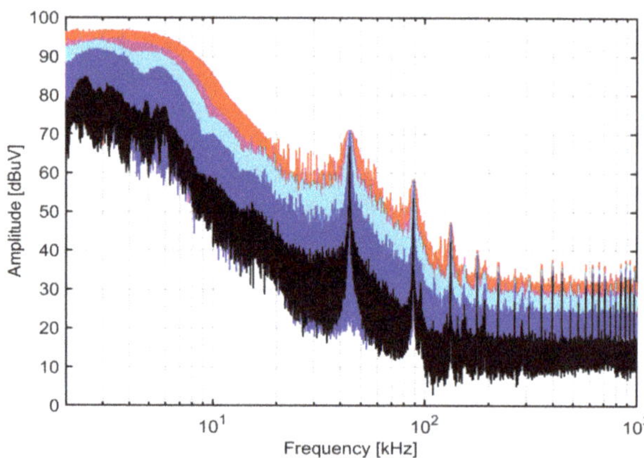

Figure 4. EMI receiver spectra with variable RBW (200 Hz, black; 1 kHz, blue; 2 kHz, light blue; 3 kHz, magenta; 5 kHz, red) for a SMPS featuring fixed switching frequency (type Black at 90% of rated power).

A slower narrow-RBW scan should be used instead (e.g., with RBW = 10 Hz or 30 Hz): the dynamic range will be maximized and, using the "max hold" setting at each frequency bin (or step), it is ensured that the repeatability of the measurement is also maximized. However, such setting is quite far from those used for measurement of emissions and the necessary scan time is extremely long (about 8 h), yielding issues of long-term stability of the setup.

3.2. *Performance of WPT*

3.2.1. Basic Time and Frequency Resolution Performance of WPT

The time resolution of both WPT spectra in Figure 5b,c is superior to that of STFT tracking the signal with 500 µs and 125 µs steps, respectively; considering the visible signal dynamics and what is offered by alternative approaches (such as a narrowband DFT or a real-time spectrum analyzer), a time step in this range is more than adequate. Correspondingly, the frequency resolution also varies by a factor of 4 in the opposite direction: when using 976 Hz in Figure 5b, the low-frequency portion of the spectrum is enriched with the details between the second and the sixth bin, which hold alternatively the maximum of the spectrum around t = 5 ms; this is evident from the asymmetry of the half-cycles of the oscillatory spike in the center of the waveform of Figure 5a. The comparison of the estimated amplitudes for the two spectra with different frequency and time resolution indicates that a finer time resolution gives slightly larger amplitudes (Figure 5c), avoiding the averaging implicit in a larger time interval. This behavior is common with the DFT and is hardly predictable quantitatively, as it depends on the characteristics of the signal components (narrowband/broadband and steady/transient). The average of four adjacent time bins in Figure 5c gives approximately the value of one bin of Figure 5b (within 1 dB). Nonetheless, the maximum value can differ by up to 3.4 dB, as occurring in the first frequency bin at 5 ms; this represents then the maximum absolute error between the two spectral representations.

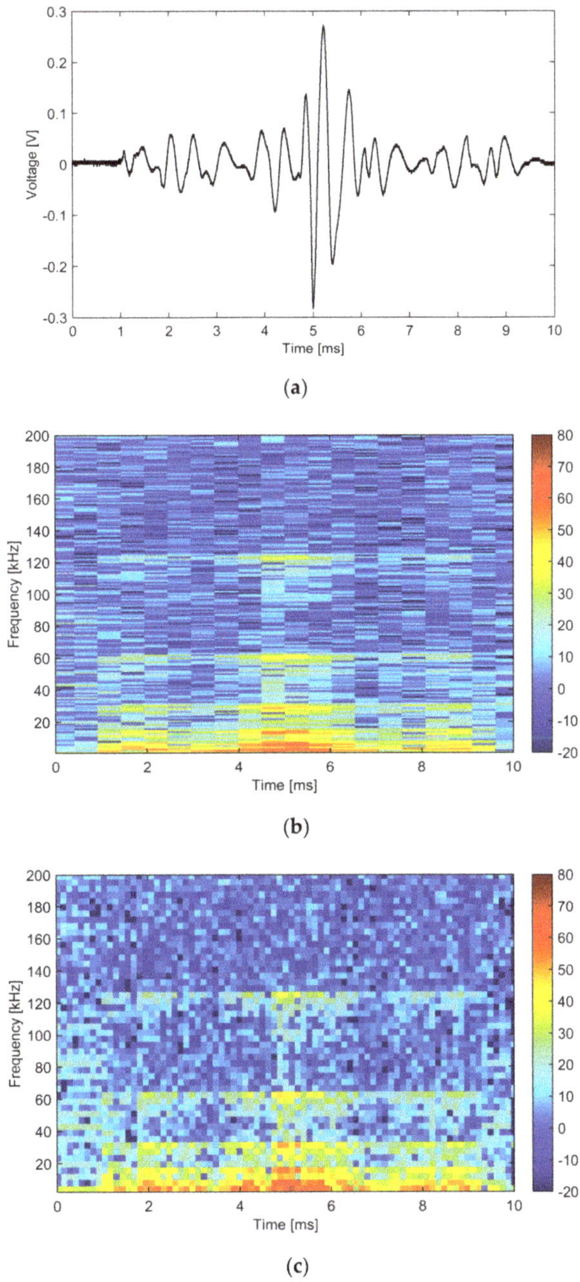

Figure 5. WPT spectrum in dBμV/Hz (normalized by frequency resolution). An example of change of dilation in time and frequency by a factor of 4 with *symlet1* wavelet: (**a**) signal waveform (10 MSa/s); (**b**) *symlet1* with $L = 9$ levels and a downsampling of $q = 10$; and (**c**) *symlet1* with $L = 8$ levels and $q = 5$. SMPS type Black at 25% of rated power.

The case of two different wavelets is shown in Figure 6, applied to the signal using the same scaling: the amplitude estimates are quite similar, with the *bior1.3* in Figure 6b, showing slightly smoother amplitude profile than the *symlet1*. Quantitatively, the average difference for the bins with the largest amplitude (those in Figure 6 with amplitude above 50 dB) is only 1.3 dB; the peaks, which are the most relevant for compliance to limits, are reproduced quite reliably with almost identical values (average difference of less than 0.25 dB). Thus, at least for the considered case, the choice of wavelets with "similar" characteristics and of suitable order for the observed time support gives repeatable results, although it represents a source of systematic error and the contribution to the uncertainty of the estimate of spectrum amplitude is not negligible (0.25 dB on average for the considered case).

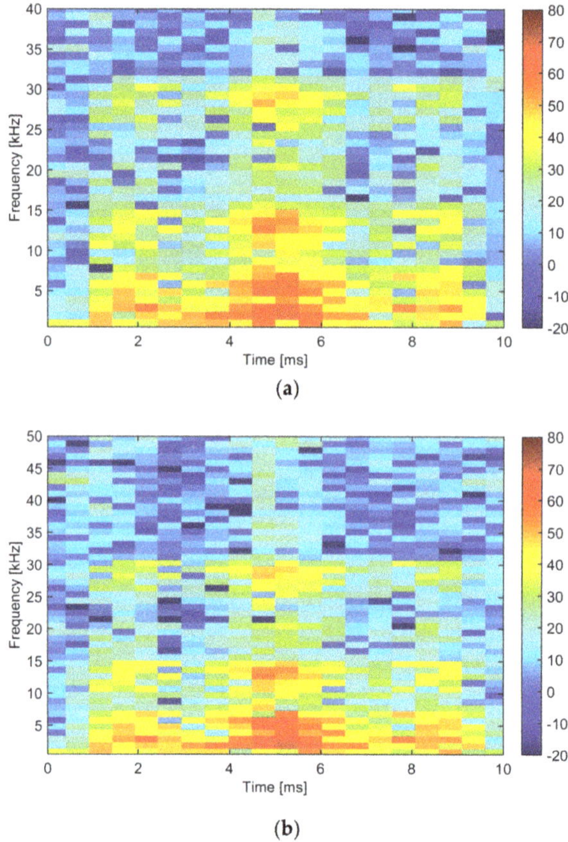

Figure 6. (**a**) Zoom on the frequency axis of Figure 5b for the frequency resolution of 976 Hz, where the change of the instantaneous frequency between 4 and 6 ms is well visible; and (**b**) additional spectrum obtained by replacing *symlet1* with *bior1.3*.

The compactness in time and frequency was evaluated considering two wavelets extensively used for PQ studies, the *db15* (equivalent to the *db20*) and the *sym8*. The results shown in Figure 7 (for the SMPS type Black at 90% of the rated power) reveal that *sym8* has some spectral leakage at the occurrence of the two peaks of the signal, at 1 ms and 4.5 ms; *db15* instead shows a limited increase of some scattered components without any visible burst of spectral leakage. Regarding the low frequency components, *db15* also shows a more coherent behavior where the second frequency bin keeps the lead for 2 ms giving yield then to fourth bin, with a return around 4 ms. Behavior of *sym8* instead is more chaotic with no definite bin prevailing for at least some ms.

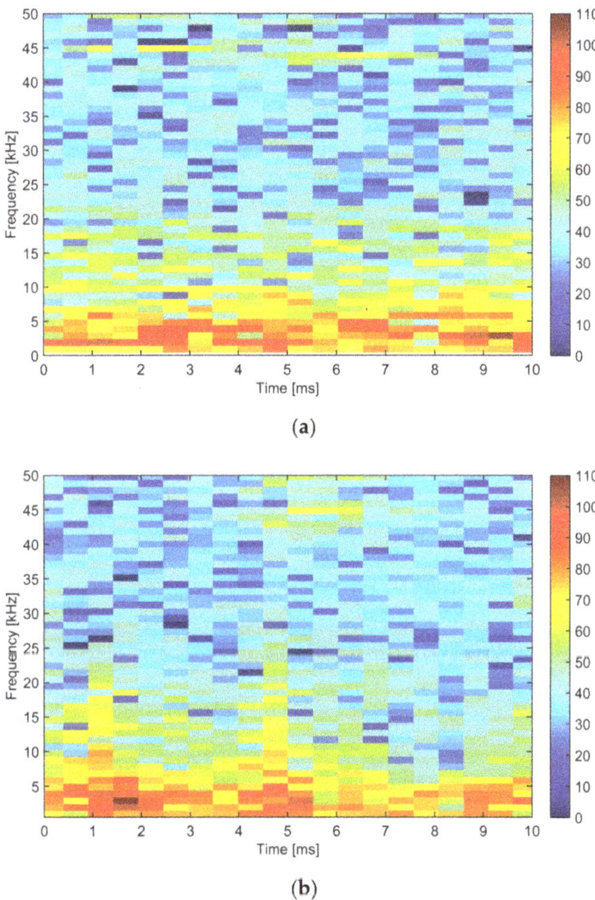

Figure 7. Comparison of two different wavelets in relation to localization on the time axis and spread to adjacent frequency bins; frequency resolution $\Delta f = 976$ Hz, number of levels $L = 9$: (**a**) *db15*; and (**b**) *sym8*. SMPS type Black at 90% of rated power.

As observed for the EEMD in Section 3.3, the best amplitude accuracy is reached for the faster components (those at about 44 and 88 kHz) that have better localization in both time and frequency: in this case, agreement for different frequency and time resolutions is within 0.6 dB for the results shown in Figure 5 (confirmed by some other tests done on different SMPSs). For nonstationary low-frequency components, the reference measurements themselves for assessment of accuracy are deemed by variability and uncertainty caused by the very nature of the signals.

3.2.2. Amplitude Accuracy and Spectrum Representation of WPT

The WPT behavior for the SMPS with constant switching frequency (named "Black" SMPS in the following) is analyzed showing the elements that concur to determine the amplitude accuracy.

As introduced in Section 2.3, the WPT spectra built with wpspectrum( ) is artificially adjusted so that the local intensity of each tile can be evaluated standalone. The adjustment consists in the multiplication by the square root of the number a of tiles in the time direction ($a = 20$ in the present case) that is re-absorbed when aggregating in the rms sense all such tiles for each frequency bin to recover its rms, which for single tone calibration signals corresponds to the rms amplitude of the test signal itself.

The results of a comparison with STFT with 1 kHz frequency resolution and calculated using the Flat Top window are shown in Figure 8.

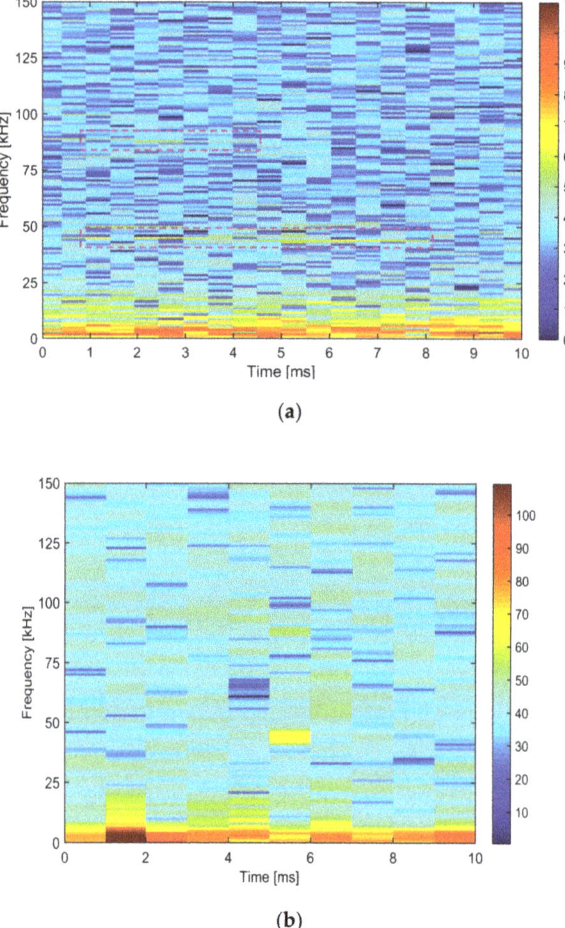

Figure 8. Black SMPS tested at 90% of rated power: (**a**) WPT spectrum obtained with *db15*, $L = 9$, $\Delta f = 976$ Hz; and (**b**) STFT spectrum with $\Delta f = 1$ kHz and time step $\Delta t = 1$ ms.

The two areas encircled by the pink rectangles are those of the 44 kHz and 88 kHz components, for which the WPT provides a thorough tracking (see Figure 8a), better than that appearing in the STFT spectrum (Figure 8b). The comparison between WPT spectrum and STFT shown in Figure 8 reveals that:

- WPT frequency resolution is superior, in terms of spectrum details and reduced leakage, which instead affect the STFT results (see, for instance, the central portion of the spectrum between 4 ms and 6 ms and the dynamic range of more than 30 dB for the background components (blue to light blue color)).
- The time resolution is also superior with the ability of tracking more closely signal dynamics.

- Regarding amplitude accuracy, we must distinguish between: (i) For narrowband switching components visible at 44 kHz and 88 kHz, there is a general agreement among WPT, STFT and the EMI receiver scan in frequency domain. We must observe that the receiver scan was made for a time interval much longer than the one covered by WPT and STFT, so that using max hold slightly larger values may be expected. (ii) The low frequency components are a byproduct of the switching pulses and evidently the instantaneous frequency is slightly variable, as result of non-linearity during oscillations. WPT confirms good tracking of such components and averaged values over adjacent bins are quite stable with respect to different frequency resolutions Δf.

Table 1 reports the results of the assessment of the amplitude accuracy for the three most prominent frequency bins shown in the first column. Reported values are the maxima, in line with the use of Flat Top for the STFT and max hold for the EMI receiver.

Table 1. Comparison of amplitudes of main spectrum components calculated with WPT and STFT and measured with EMI receiver (using the 1 kHz RBW results in Figure 4).

Frequency (kHz)	EMI Rec. (dBμV)	WPT (dBμV)	STFT (dBμV)
2–5	93.0	93.75–105.5 (1)	82.73–106.7 (1)
44	70.5	68.07, 71.51 (2)	69.04
88	57.5	60.60 (3)	53.27

(1) Range of maxima in the 2–4 kHz bins over 10 ms for 90% of bins; implicit averaging in this case is extremely important and WPT has the best resolution and the least averaging. (2) Five bins between 60.64 dBμV and 65.87 dBμV; with rms summation of adjacent bins carried out to cope with the 1 ms time resolution of the STFT, the results are 71.51 dBμV and 68.07 dBμV centered at 2.5 ms and 6.5 ms. (3) Two bins with 55.27 dBμV and 53.83 dBμV; with rms summation of adjacent bins carried out to cope with the 1 ms time resolution of the STFT, the result is 60.60 dBμV centered at 2.5 ms.

3.3. EMD and EEMD Performance

The EMD and EEMD were used in the R interface implementation Rlibeemd [30,31]. The measured voltage is analyzed first using EMD. As the boundaries of the signal may affect the decomposition of EMD-based algorithms [32], to avoid boundary problems, particular care was taken in the choice of an appropriate time record of 10 ms. Figure 9 shows the 12 IMFs extracted from the original signal. As demonstrated in [33], the EMD performs as a filter-bank sifting out the high-frequency harmonics first, whereas the low-frequency ones pass through the filter. IMFs 2 and 3 contain the information needed to represent the significant part of the frequency spectrum, i.e., the three peaks corresponding to the switching frequency (about 44 kHz) of the SMPS and its second and third harmonics. IMF 2 contains the information related to the second and third harmonics of the fundamental frequency, whereas IMF 3 contains the information of the switching frequency. Performing a FFT on the sum of these two IMFs yields the red spectrum in Figure 10, whereas the blue spectrum is the FFT of the original signal. A Flat Top window is applied to the original signal and IMFs in order to maximize the amplitude estimate [15].

As shown in Figure 10, the three peaks are predicted with good accuracy, although there is a difference of about 1 dB with the FFT of the original signal for the peak at the fundamental frequency. Considering these two IMFs 2 and 3 only has the effect of filtering the low-frequency content of the measured voltage, which is collected in IMFs 4–13. IMFs 4 and 5 contain the information related to the 100 Hz ripple components (whose period corresponds to 10 ms, as shown in Figure 2). These IMFs are those with the largest amplitude and represent the low-frequency part of the signal in the frequency domain. IMFs 2–5 thus contain sufficient information to represent the signal, as shown in Figure 11. As can be seen, this spectrum slightly differs from the spectrum of the original signal up to 150 kHz. The fundamental, second and third harmonics are predicted with good accuracy with respect to the FFT of the original signal. Some information on the fundamental frequency may then be contained in IMFs 4 and 5 due to mode mixing, as mentioned in Section 2.4. Of course, the upper part of the frequency range is not predicted accurately, as the lowest order IMF (that containing the information on higher frequencies) has not been considered. Nevertheless, the supraharmonics spectrum 2–150 kHz is well represented with just four IMFs obtained with EMD.

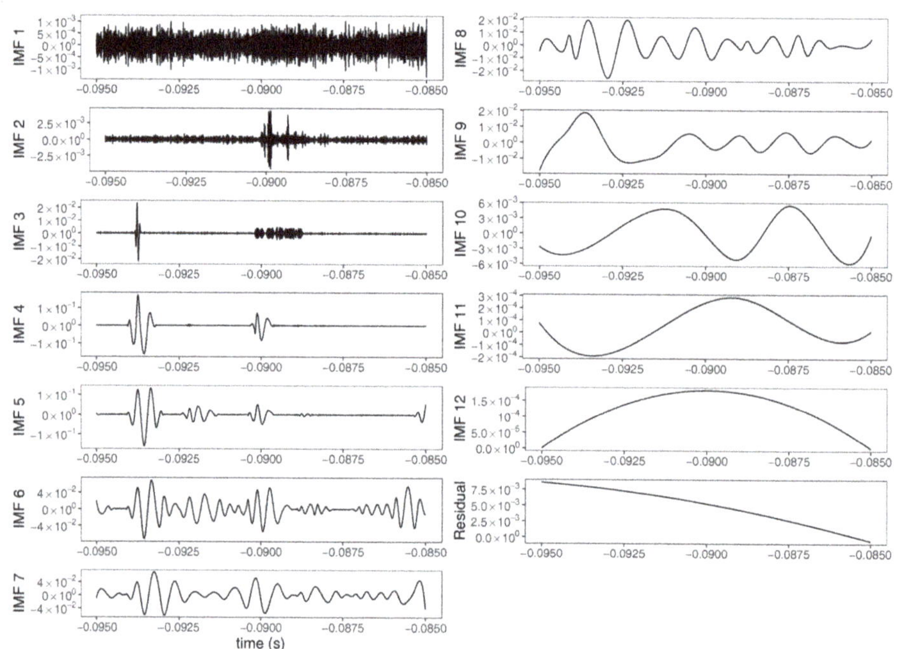

Figure 9. IMFs extracted with the EMD procedure.

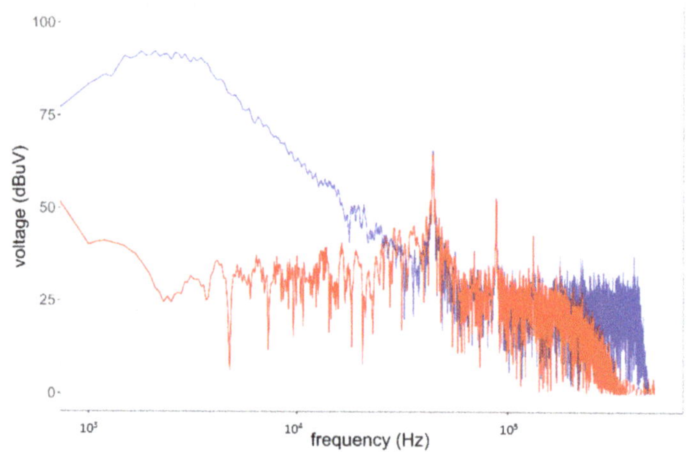

Figure 10. Comparison of the FFT spectra obtained from IMFs 2 and 3 extracted with EMD (**red**) and from the original signal (**blue**).

The application of the EEMD yields the same number of IMFs as EMD, 12, as shown in Figure 12. The information related to the 100 Hz ripple components is contained in IMFs 5–9, being IMFs 6 and 7 those with the largest amplitude. To represent the signal in the frequency domain, only IMFs 2–7 are needed. The high-frequency information of the signal is now distributed over three IMFs (IMFs 2–4), as shown in Figure 13. The accuracy with which the first two peaks are determined is very good; however, although the amplitude of the third harmonic of the fundamental is predicted quite accurately, the noise floor beyond 100 kHz raises making it less clearly visible than in the case of the EMD. The filtering effect on the low-frequency components of the spectrum is still evident. By adding IMFs 5–7 and performing the FFT on the sum of these six IMFs, the red spectrum of Figure 14 is obtained, which is in fairly good agreement with that of the original signal.

By means of the application of FFT to the relevant modal components obtained with EMD, it is shown that the frequency extraction ability of the EMD is consistent and strong [34].

The results of comparison with STFT with 1 kHz frequency resolution, a 1 ms time step and a Flat Top window are shown in Figures 15 and 16. The former figure refers to the application of STFT to the signal obtained composing IMFs 2 and 3 extracted with EMD; the latter to the application of STFT to the signal obtained composing IMFs 2–4 extracted with EEMD. The amplitudes are the maximum values obtained by the successive FFTs that compose the STFT, as it would be with a spectrum analyzer set to max hold. It can be noticed that the amplitudes of the first peak (at 44 kHz, with 68.27 dBμV and 69.97 dBμV, respectively) and of the second peak (at 88 kHz, with 57.59 dBμV and 58.51 dBμV, respectively) are in agreement with those shown in Table 1 for WPT and EMI receiver data, thus confirming the suitability of the approach in the analysis of EMI.

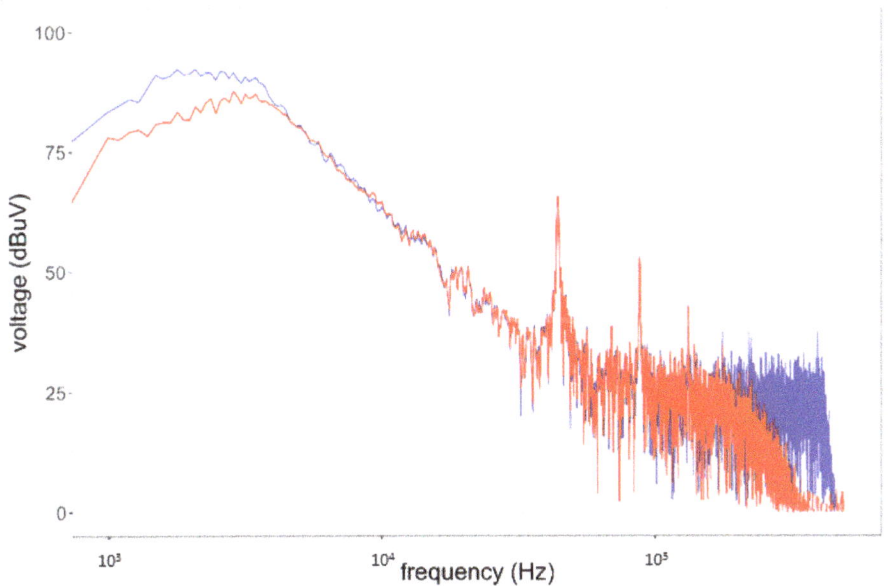

Figure 11. Comparison of the FFT spectra obtained from IMFs 2–5 extracted with EMD (**red**) and from the original signal (**blue**).

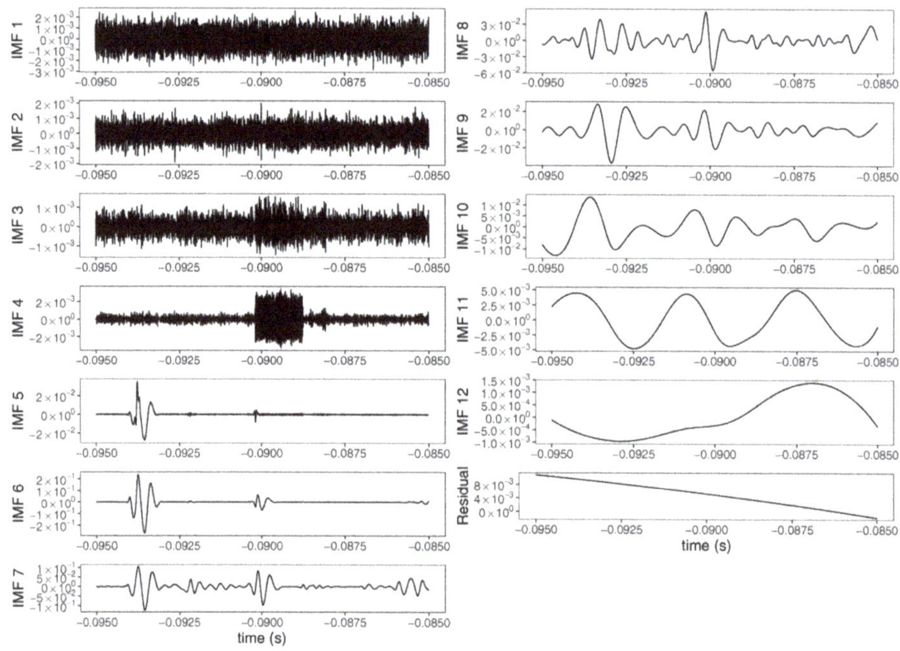

Figure 12. IMFs extracted with the EEMD procedure.

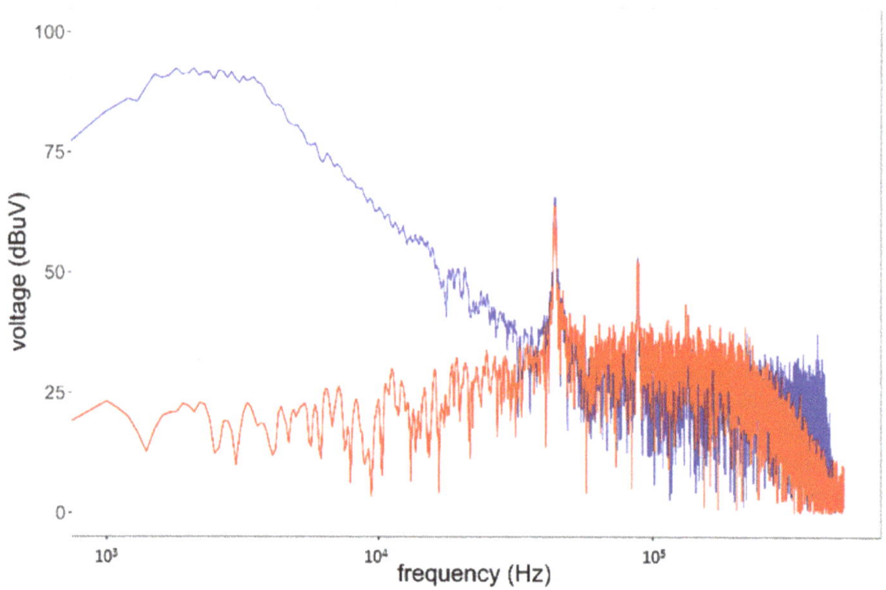

Figure 13. Comparison of the FFT spectra obtained from IMFs 2–4 extracted with EEMD (**red**) and from the original signal (**blue**).

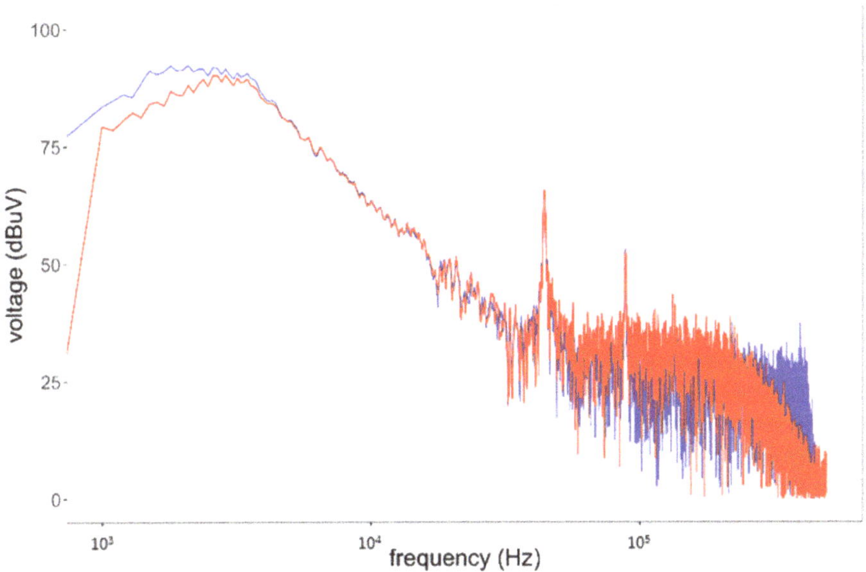

Figure 14. Comparison of the FFT spectra obtained from IMFs 2–7 extracted with EEMD (**red**) and from the original signal (**blue**).

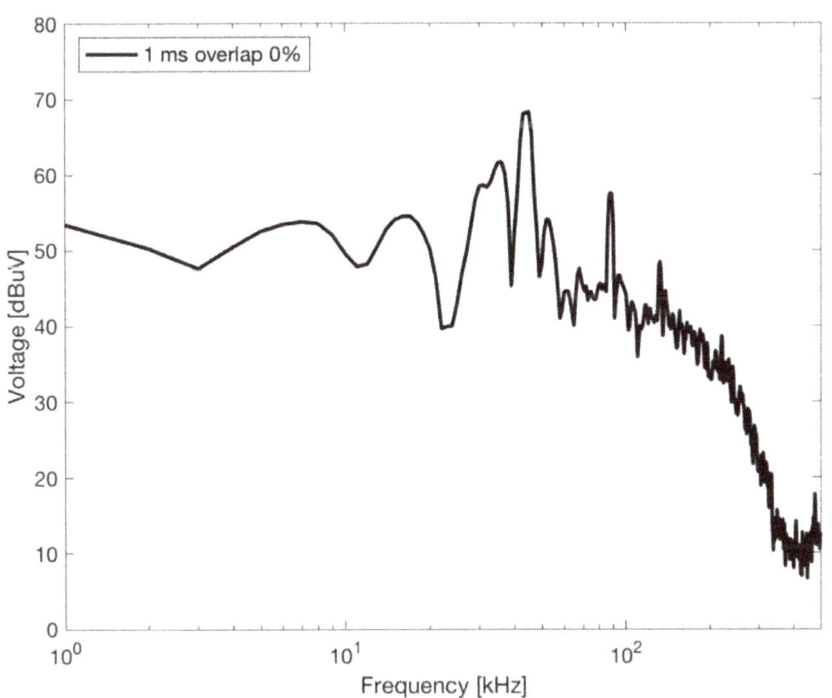

Figure 15. STFT spectrum with $\Delta f = 1$ kHz and time step $\Delta t = 1$ ms of the signal resulting from the composition of IMFs 2 and 3 extracted with EMD.

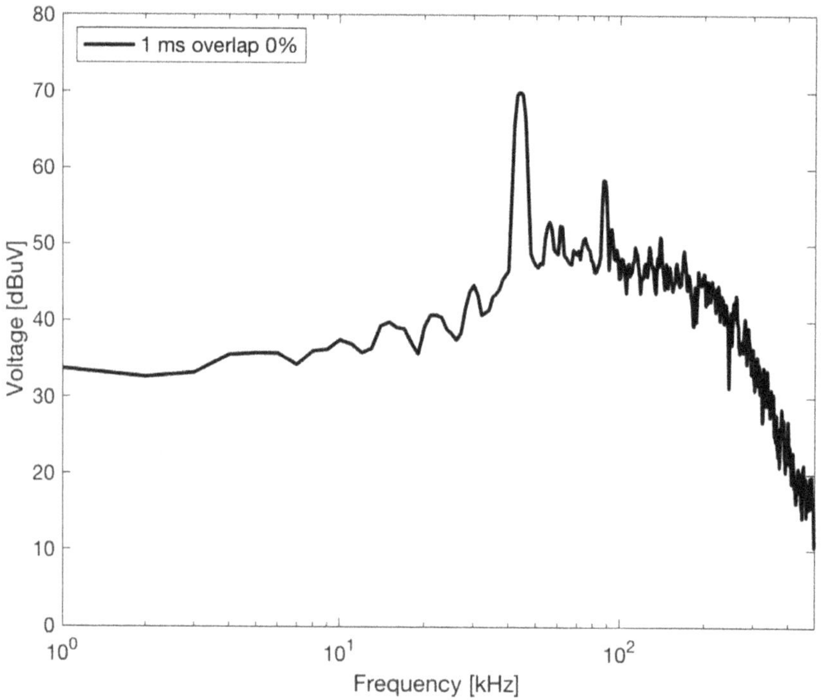

Figure 16. STFT spectrum with Δf = 1 kHz and time step Δt = 1 ms of the signal resulting from the composition of IMFs 2 to 4 extracted with EEMD.

4. Conclusions

The problem of the assessment of conducted emissions of Switched Mode Power Supplies (SMPSs) using time-domain sampled data is considered. Emissions measured with a Digital Sampling Oscilloscope using a Line Impedance Stabilization Network are characterized by fast switching pulses and damped oscillations, posing a problem for reconciling amplitude accuracy, suitable frequency resolution and fast time response to track signal dynamics.

Suitable processing techniques are discussed, contrasted to the most popular approach of using Discrete Fourier Transform analysis. The DFT drawbacks are pointed out in Section 2, in relation to the unfavorable time support and frequency resolution, as well as to the known issues of amplitude accuracy and spectral leakage. Two techniques (Wavelet Packet Transform (WPT) and Empirical Mode Decomposition (EMD)) based on different principles are then evaluated, with the objective of an adequate and accurate handling of signal characteristics, as reflected in its spectral components and their amplitude, for the assessment of SMPS conducted emissions.

WPT and EMD are compared for complexity, ease of implementation and, most of all, usability of the output and suitability to identify and represent the relevant spectral components. The objective is that of an accurate amplitude estimate and assessment of hypothetical compliance to limits of emissions. The output representation of the WPT is more effective, in that the data are represented as one might expect for not only comparison with limits, but also analysis of short-term disturbance (e.g., for victim circuits characterized by some transient susceptibility) and in general tracking and troubleshooting of byproducts along the evolution of the signal.

WPT shows a general consistency and robustness with respect to the change of the number of levels L and downsampling q (applied to the original 10 MSa/s sample rate). Similarly, the change of mother wavelet is verified, passing in one case from the *symlet* (linear phase and asymmetric) to the *biorthogonal* (linear phase and symmetric), using for each the minimum order, and in a second case using db15 ad sym8, observing a larger spectrum leakage for the latter. The sensitivity analysis indicates a variability of less than 1 dB when comparing values averaged with the adjacent time and frequency bins; maximum observed difference for the peak values is 3.6 dB for the broad 2.2 kHz component that suffers the largest variability.

The application of EMD and EEMD algorithms yields a limited number of IMFs that are sufficient to represent the original signal fairly accurately (four and six, respectively). Moreover, the representation can be made even more compact by exploiting the filtering effect of the IMFs on the low-frequency components of the signal, thus reducing the number of IMFs necessary to represent the most significant part of the spectrum to two and three, respectively. EMD represents the spectrum for frequencies larger than 100 kHz better than EEMD, which seems to produce a higher noise floor.

Future work will be focused on the identification of quantitative indices of performance and a framework to evaluate systematically the performances of the various implementations, especially for the WPT, where the number of mother wavelets and filter implementations is almost uncountable.

Author Contributions: The contribution of L.S. and A.M. to conceptualization, methodology, software, experimental validation, and writing is the same. All authors have read and agreed to the published version of the manuscript.

Funding: This research received no external funding.

Conflicts of Interest: The authors declare no conflict of interest.

References

1. Thomas, D. Conducted emissions in distribution systems (1 kHz–1 MHz). *IEEE Electromagn. Compat. Mag.* **2013**, *2*, 101–104. [CrossRef]
2. Larsson, E.A.; Bollen, M.H.; Wahlberg, M.G.; Lundmark, C.M.; Ronnberg, S.K. Measurements of high-frequency (2–150 kHz) distortion in low-voltage networks. *IEEE Trans. Power Deliv.* **2010**, *25*, 1749–1757. [CrossRef]
3. Frey, D.; Schanen, J.; Quintana, S.; Bollen, M.; Conrath, C. Study of high frequency harmonics propagation in industrial networks. In Proceedings of the EMC Europe, Rome, Italy, 17–21 September 2012.
4. Bartak, G.F.; Abart, A. EMI in the frequency range 2 kHz–150 kHz. In Proceedings of the 22nd International Conference and Exhibition on Electricity Distribution (CIRED 2013), Stockholm, Sweden, 10–13 June 2013.
5. Klatt, M.; Meyer, J.; Schegner, P. Comparison of measurement methods for the frequency range of 2 kHz to 150 kHz. In Proceedings of the 2014 16th International Conference on Harmonics and Quality of Power (ICHQP), Bucharest, Romania, 25–28 May 2014.
6. Sandrolini, L.; Thomas, D.W.P.; Sumner, M.; Rose, C. Measurement and Evaluation of the Conducted Emissions of a DC/DC Power Converter in the Frequency Range 2–150 kHz. In Proceedings of the 2018 IEEE Symposium on Electromagnetic Compatibility, Signal Integrity and Power Integrity (EMC, SI & PI), Long Beach, CA, USA, 30 July–3 August 2018.
7. Sandrolini, L.; Mariscotti, A. Techniques for the Analysis of Time-Domain Conducted Emissions of SMPS in Smart Grids. In Proceedings of the 2019 IEEE International Conference on Environment and Electrical Engineering and 2019 IEEE Industrial and Commercial Power Systems Europe, Genova, Italy, 11–14 June 2019.
8. Ensini, L.; Sandrolini, L.; Thomas DW, P.; Sumner, M.; Rose, C. Conducted Emissions on DC Power Grids. In Proceedings of the International Symposium on Electromagnetic Compatibility (EMC EUROPE), Amsterdam, The Netherlands, 27–30 August 2018.
9. Sandrolini, L.; Pasini, G. Study of the Conducted Emissions of an SMPS Power Converter from 2 to 150 kHz. In Proceedings of the IEEE International Conference on Environment and Electrical Engineering and 2018 IEEE Industrial and Commercial Power Systems Europe, Palermo, Italy, 12–15 June 2018.
10. CENELEC EN 61000-4-19. *Electromagnetic Compatibility (EMC)—Part 4–19: Testing and Measurement Techniques—Test for Immunity to Conducted, Differential Mode Disturbances and Signalling in the Frequency Range 2 kHz to 150 kHz at a.c. Power Ports*; European Committee for Electrotechnical Standardization: Brussels, Belgium, 2014.

11. CENELEC EN 55014-1. *Electromagnetic Compatibility—Requirements for House—Hold Appliances, Electric Tools and Similar Apparatus—Part 1: Emission*; European Committee for Electrotechnical Standardization: Brussels, Belgium, 2017.
12. CENELEC EN 55015. *Limits and Methods of Measurement of Radio Disturbance Characteristics of Electrical Lighting and Similar Equipment*; European Committee for Electrotechnical Standardization: Brussels, Belgium, 2014.
13. CENELEC EN 61000-4-30. *Electromagnetic Compatibility—Part 4–30: Testing and Measurement Techniques—Power Quality Measurement Methods*; European Committee for Electrotechnical Standardization: Brussels, Belgium, 2015.
14. CENELEC EN 50065-1. *Signalling on Low-Voltage Electrical Installations in the Frequency Range 3 kHz to 148.5 kHz—Part 1: General Requirements, Frequency Bands and Electromagnetic Disturbances*; European Committee for Electrotechnical Standardization: Brussels, Belgium, 2012.
15. Sandrolini, L.; Mariscotti, A. Impact of Short-Time Fourier Transform Parameters on the Accuracy of EMI Spectra Estimates in the 2–150 kHz Supraharmonics Interval. *Electr. Power Syst. Res.*. under review.
16. Barros, J.; Diego, R.I. Analysis of Harmonics in Power Systems Using the Wavelet-Packet Transform. *IEEE Trans. Instrum. Meas.* **2008**, *57*, 63–69. [CrossRef]
17. Masoum, M.A.S.; Jamali, S.; Ghaffarzadeh, N. Detection and classification of power quality disturbances using discrete wavelet transform and wavelet networks. *IET Sci. Meas. Technol.* **2010**, *4*, 193–205. [CrossRef]
18. Lodetti, S.; Bruna, J.; Melero, J.J.; Sanz, J.F. Wavelet Packet Decomposition for IEC Compliant Assessment of Harmonics under Stationary and Fluctuating Conditions. *Energies* **2019**, *12*, 4389. [CrossRef]
19. Zhu, S.-X.; Liu, H. Simulation study of power harmonic based on Daubechies wavelet. In Proceedings of the 2010 International Conference on E-Product E-Service and E-Entertainment, Henan, China, 7–9 November 2010.
20. Mariscotti, A. Characterization of Power Quality transient phenomena of DC railway traction supply. *Acta Imeko* **2012**, *1*, 26–35. [CrossRef]
21. Mariscotti, A. Methods for Ripple Index evaluation in DC Low Voltage Distribution Networks. In Proceedings of the IEEE Instrumentation and Measurement Technical Conference (IMTC), Warsaw, Poland, 2–4 May 2007.
22. Hai, Y.; Chen, J.-Y. RMS and power measurements based on wavelet packet decomposition using special IIR filter bank. In Proceedings of the 2011 International Conference on Electronic & Mechanical Engineering and Information Technology, Harbin, China, 12–14 August 2011.
23. Huang, N.E.; Wu, Z. A review on Hilbert-Huang transform: Method and its applications to geophysical studies. *Rev. Geophys.* **2008**, *46*, 1–23. [CrossRef]
24. Li, H.; Zhang, Y.; Zheng, H. Hilbert-Huang transform and marginal spectrum for detection and diagnosis of localized defects in roller bearings. *J. Mech. Sci. Technol.* **2009**, *23*, 291–301. [CrossRef]
25. Fu, K.; Qu, J.; Chai, Y.; Zou, T. Hilbert marginal spectrum analysis for automatic seizure detection in EEG signals. *Biomed. Signal Proc. Control* **2015**, *18*, 179–185. [CrossRef]
26. Zhaohua, W.; Huang, N.E. Ensemble Empirical Mode Decomposition: A Noise-Assisted Data Analysis Method. *Adv. Adapt. Data Anal.* **2009**, *1*, 1–41.
27. Gong, X.; Ferreira, J.A. Comparison and Reduction of Conducted EMI in SiC JFET and Si IGBT-Based Motor Drives. *IEEE Trans. Power Electron.* **2014**, *29*, 1757–1767. [CrossRef]
28. Mariscotti, A. Normative Framework for the Assessment of the Radiated Electromagnetic Emissions from Traction Power Supply and Rolling Stock. In Proceedings of the 2019 IEEE Vehicle Power and Propulsion Conference (VPPC), Hanoi, Vietnam, 14–17 October 2019.
29. IEC 61000-4-7. *Electromagnetic Compatibility (EMC)—Part 4–7: Testing and Measurement Techniques—General Guide on Harmonics and Interharmonics Measurements and Instrumentation, for Power Supply Systems and Equipment Connected Thereto*; International Electrotechnical Commission: Geneva, Switzerland, 2009.
30. Helske, J.; Luukko, P.J. Rlibeemd: Ensemble Empirical Mode Decomposition (EEMD) and Its Complete Variant (CEEMDAN); R Package Version 1.4.1. 2018. Available online: https://github.com/helske/Rlibeemd (accessed on 1 October 2020).
31. Luukko, P.J.; Helske, J.; Räsänen, E. Introducing libeemd: A program package for performing the ensemble empirical mode decomposition. *Comput. Stat.* **2006**, *31*, 545–557. [CrossRef]
32. Stallone, A.; Cicone, A.; Materassi, M. New insights and best practices for the successful use of Empirical Mode Decomposition, Iterative Filtering and derived algorithms. *Sci. Rep.* **2020**, *10*, 15161. [CrossRef] [PubMed]

33. Feldman, M. Analytical basics of the EMD: Two harmonics decomposition. *Mech. Syst. Signal Process.* **2009**, *23*, 2059–2071. [CrossRef]
34. Ge, H.; Chen, G.; Yu, H.; Chen, H.; An, F. Theoretical Analysis of Empirical Mode Decomposition. *Symmetry* **2018**, *10*, 623. [CrossRef]

Article

Modeling and Optimization of Impedance Balancing Technique for Common Mode Noise Attenuation in DC-DC Boost Converters

Shuaitao Zhang [1,*], Baihua Zhang [2], Qiang Lin [1], Eiji Takegami [3], Masahito Shoyama [1] and Gamal M. Dousoky [1,4,5]

1 Graduate School of Information Science and Electrical Engineering, Kyushu University, Fukuoka 819-0395, Japan; lin@ckt.ees.kyushu-u.ac.jp (Q.L.); shoyama@ees.kyushu-u.ac.jp (M.S.); dousoky@mu.edu.eg (G.M.D.)
2 Institute of Engineering Thermophysics, China Academy of Science, Beijing 100190, China; 09291159@bjtu.edu.cn
3 Technology Development Department, TDK-Lambda Corporation, Niigata 940-1195, Japan; e.takegami@jp.tdk-lambda.com
4 Electrical Engineering Department, Minia University, Minia 61517, Egypt
5 Department of Communications and Computer Engineering, Nahda University, New Beni Suef 62513, Egypt
* Correspondence: szhang@ckt.ees.kyushu-u.ac.jp

Received: 13 February 2020; Accepted: 13 March 2020; Published: 14 March 2020

Abstract: As an effective means of suppressing electromagnetic interference (EMI) noise, the impedance balancing technique has been adopted in the literature. By suppressing the noise source, this technique can theoretically reduce the noise to zero. Nevertheless, its effect is limited in practice and also suffers from noise spikes. Therefore, this paper introduces an accurate frequency modeling method to investigate the attenuation degree of noise source and redesign the impedance selection accordingly in order to improve the noise reduction capability. Based on a conventional boost converter, the common mode (CM) noise model was built by identifying the noise source and propagation paths at first. Then the noise source model was extracted through capturing the switching voltage waveform in time domain and then calculating its Fourier series in frequency domain. After that, the conventional boost converter was modified with the known impedance balancing techniques. This balanced circuit was analyzed with the introduced modeling method, and the equivalent noise source was precisely estimated by combining the noise spectra and impedance information. Furthermore, two optimized schemes with redesigned impedances were proposed to deal with the resonance problem. A hardware circuit was designed and built to experimentally validate the proposed concepts. The experimental results demonstrate the feasibility and effectiveness of the proposed schemes.

Keywords: common mode noise; EMI modeling; Fourier series; impedance balancing; resonant frequency

1. Introduction

The switch-mode power supply (SMPS) has played an important role in power generation and conversion systems, which tend to require a higher reliability, a higher power density, and a longer working life. However, the controllable switches adopted in SMPS, like metal-oxide-semiconductor field-effect transistors (MOSFETs), insulated gate bipolar transistors (IGBTs) and especially, the newest wide bandgap (WBG) semiconductor switches, may generate violent voltage and current gradients (dv/dt and di/dt). Those gradients bring a serious EMI problem and threaten the normal operation of electrical equipment. Typically, relevant electromagnetic compatibility (EMC) standards which include conducted emission and radiated emission must be fulfilled before the products enter the market [1,2].

Plenty of efforts have been made to investigate the generation and propagation mechanisms of the conducted noise. Among these investigation approaches, establishing an EMI noise model helps to predict noise spectrum results and adopt appropriate attenuation strategies in the early stage [3–15]. The lumped circuit modeling method, which depends on the time-domain circuit simulation has been introduced in [3,4]. This general model contains the coefficients of each element and then give EMI prediction result by utilizing simulation software. This model exactly corresponds to the actual circuit, with traits of intuitive and easy to understand. The results comparison shows that the established noise model gives an accurate prediction of conducted noise up to several MHz. However, the extraction of detailed parasitic parameters is difficult in practice, not to mention that the simulation process is time-consuming and may encounter convergence problems.

To process faster and get stable predictions, the frequency domain approach has been developed [5–15]. The early-stage modeling methods used simplified lumped parameters and ideal geometric shapes to replace the circuit impedance and the noise source, respectively [5,6]. However, such approximation has to deal with over-simplification and the accuracy of noise prediction needs further improvement.

Recently, a new branch of frequency domain modeling, namely the "behavioral modeling method" has been proposed and received extensive attention [7–9]. Instead of establishing the common (CM) noise model and differential mode (DM) noise model respectively, the behavioral model could predict the CM and DM noise in one integrated model. Both the two-terminal (one port) model and three-terminal (two ports) model have been extracted based on Thevenin or Norton theorems. This method would work well when the object topology is not very complicated, like non-isolated converters and half-bridge circuits. Nevertheless, without the analysis of noise propagation, the impedance identification has always troubled by the lack of a physical meaning. This problem becomes especially significant when modeling complex circuits. The impedance error would continue to affect the precision of the prediction results. Even repeating measurements under different series and shunt conditions may not help to reduce the error.

In order to do further research on the noise generation mechanism and improve the accuracy of the frequency domain model, some works have been focused on the switch voltage waveform [10–15]. Generally speaking, the switch has been substituted by a voltage or current source to represent the electric potential fluctuation. By replacing the switch waveform with the trapezoid, the simulation process was simplified and the predicted conducted emissions could match the experimental results around 3MHz [10]. Nevertheless, the fact that the real voltage waveform is not exactly a standard trapezoid should not be neglected. The authors of [11–15] have studied the switching dynamic process of the semiconductor in depth. More details have been concluded in the noise source model and thus help to improve the accuracy of the predicted results.

As an effective method to reduce the conducted noise, the impedance balancing technique has been discussed and implemented in many articles [16–21]. In this technique, the switch is surrounded by the impedance bridges while the impedance ratios of each two bridge arms remain the same. Therefore, the noise source could be greatly attenuated and the explanation can stem from the Wheatstone bridge theory. This technique can be flexibly applied to different circuits by adjusting the propagation impedance. The conventional boost converter (CBC) is handled so that the power inductor is spilled into two parts and the parasitic capacitances are adjusted. Furthermore, in the case of inverters, the general method is to coordinate with the external CM choke circuit [18]. The noise level has been suppressed well in these situations, but there are still some problems. Firstly, a serious noise peak occurred because of the impedance resonance. Moreover, it lacks an accurate model that could effectively estimate attenuation effects.

In this work, a novel CM noise modeling method was introduced to analyze and optimize the impedance balancing technique. Specifically, the attenuation degree of noise source was investigated by comparing the impedance and spectra information, and the resonance problem was solved effectively by redesigning the bridge impedances. Since the application of balance would not affect the operation of the switch, the CBC was modeled at first. The accurate noise source method plays a vital role in improving the accuracy of the overall model.

This paper is organized as follows. Section 2 introduces the CM noise model of CBC and gives details of noise source modeling. Section 3 analyzes the mechanism of balancing impedance technique and evaluates the results in the view of impedance and equivalent noise source. Section 4 proposes two schemes to solve the resonance problem and gives experimental results to certify their validity. Finally, Section 5 provides conclusions of this work.

2. Formulation of the Modeling Method with an Accurate Noise Source

Figure 1 shows the configuration of a conventional boost converter connecting with the Line Impedance Stabilization Network (LISN). The LISN is placed between the DC power supply and the boost converter circuit, which provides a standard impedance to the equipment under test (EUT) and evaluates the conducted noise level by integrating with a spectrum analyzer. P and N are the positive pole and negative pole of converter input and G is the frame ground of circuit and LISN. In CBC, C_i and C_o are the input side and output side capacitances, and L is the power inductor. C_g is the capacitor that connects the ground and the drain of the MOSFET switch, which simulates the parasitic capacitors between the switch and the heatsink as well as the capacitors between the heatsink and the frame ground.

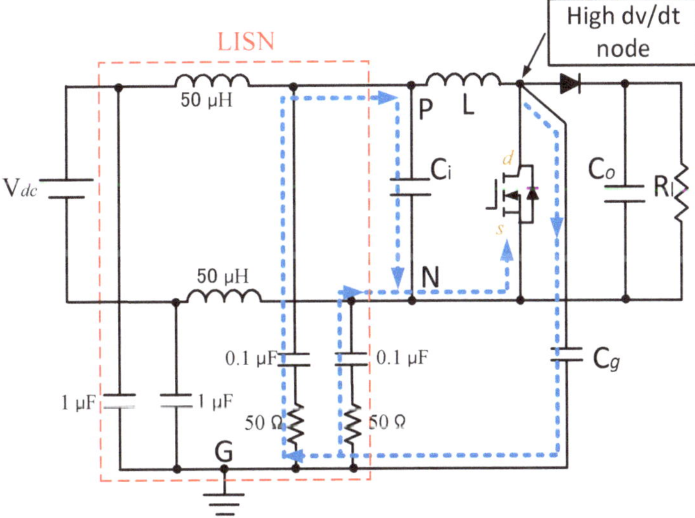

Figure 1. Configuration of the conventional boost circuit (CBC) and LISN setup.

2.1. Propagation Path of CM Noise Current and the Equivalent Model

At the conducted noise frequency band (from 150 kHz to 30 MHz), the relative impedance of the capacitor is small and the inductor is large. Therefore, C_i (100 μF) and L (500 μH) could be considered as short and open, respectively, in the CM noise model. The conducted noise was generated by the high-frequency voltage fluctuation between the drain and source of the switch. Here in the CBC, the source voltage is almost constant even when the switch is working. On the other hand, as marked as the high dv/dt node in Figure 1, the drain voltage changes rapidly and large CM current flows through C_g to the ground. In conclusion, the blue dotted line in Figure 1 indicates the flow direction of CM current.

The CM noise model resulted from the developed analysis is shown in Figure 2. Here, Z_{LISN} represents the equivalent resistance of LISN, which could be taken as a 25 Ω resistor in CM noise calculation. V_N is the noise source, represents the voltage between drain and source of the switch (namely V_{ds}). Z_{Cg} is the impedance of C_g. According to the EMI measurement specification, the length of the connecting wire between EUT and LISN should be equal to 80 cm. Here, to improve the accuracy of the noise model, the impedance of the connecting wire is considered and represented as Z_W.

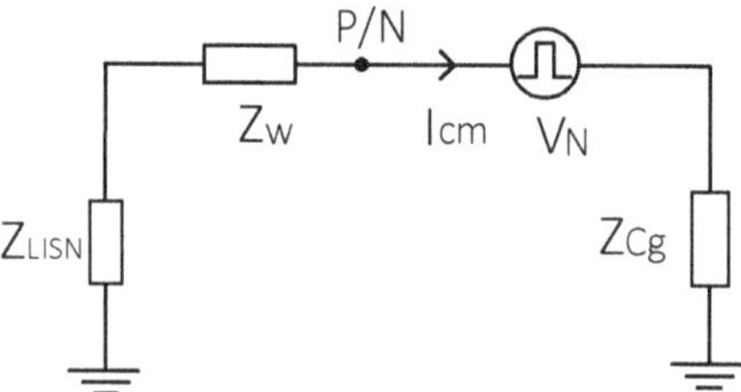

Figure 2. Common mode (CM) noise model of the conventional boost circuit.

According to the equivalent noise model, the CM noise voltage of CBC could be represented by Equation (1):

$$V_{LISN} = \frac{Z_{LISN}}{Z_{LISN} + Z_w + Z_{Cg}} V_N \tag{1}$$

This equation is true in both time and frequency domains, and the calculations in this paper are performed in a frequency domain.

2.2. Accurate Noise Source Modeling Method

As mentioned in the introduction, the precision of the noise model is largely affected by the noise source. Thus, plenty of methods have been proposed to obtain the noise source in time or in frequency domain. In this article, a novel frequency domain modeling method is introduced. The computational complexity is reduced while the accuracy is guaranteed. Specifically, the switch voltage waveform during experiment was observed and classified in time domain, and then, the overall spectrum could be obtained in frequency domain by calculating the Fourier series by parts. The accuracy of the noise source model is improved remarkably after considering the high frequency ringing.

The general steps for noise source modeling include:

(1) Capture the turning on and turning off waveforms of the switch by the oscilloscope, while adjusting the appropriate time scale in order to observe the parameters.

(2) Divide the overall waveform into three parts: trapezoid, turn-off ringing, and turn-on ringing. Extract the necessary parameters for the Fourier transform.

(3) Calculate Fourier series of each component separately and then add them together. Finally, the overall spectrum could be obtained.

To specify the modeling process, an experimental setup was established according to Figure 1. Details of the experimental configuration are given in the following section. The key parameters of the circuit are given in Table 1. In addition, 1nF Y-capacitor was adopted as C_g.

Table 1. Key parameters of CBC.

Parameter	Value
Input voltage	12 V
Switch frequency	103 KHz
Duty ratio	0.75
Load	50 Ω

Figure 3 shows the voltage waveform of switch measured by oscilloscope. The ringing caused by turn-on and turn-off could be seen clearly from the following magnified images. Note that the turn-on ringing is quite limited and thus, its effectiveness could be ignored.

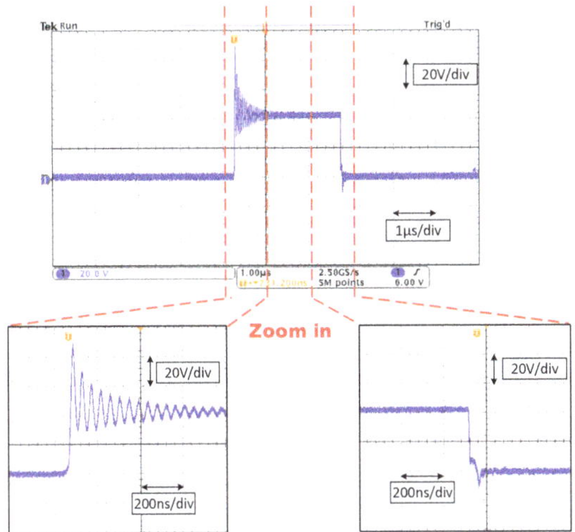

Figure 3. The experimental voltage waveform of switch measured by oscilloscope.

As the real voltage waveform is an irregular shape, it is difficult to calculate its spectrum directly. This paper decomposed the switch voltage waveform and the Fourier series could be done in steps. The corresponding decomposition process of the overall waveform is shown in Figure 4. The overall waveform is shown at the top of the figure, and it can be divided into the trapezoid component and the ringing component. This decomposition has reduced the complexity of Fourier transformation and improved the precision at the same time. The relative parameters are also marked in Figure 4, which could help to identify the corresponding symbols listed in Table 2.

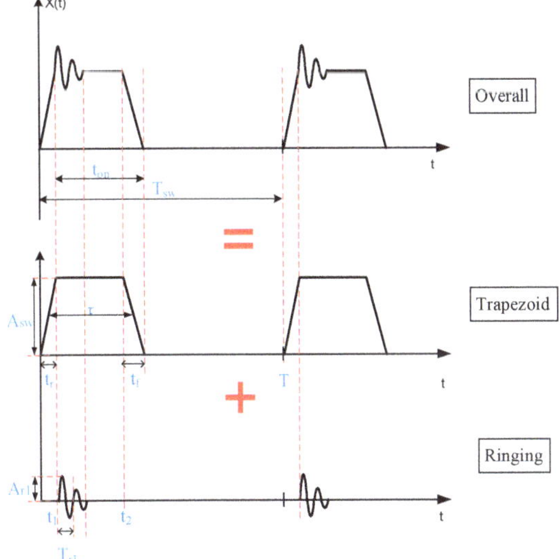

Figure 4. Decomposition diagram of the switch voltage waveform.

Table 2. Characteristic parameters of the switch voltage waveform.

Parameter	Symbol	Value
Amplitude of trapezoid	A_{sw}	42.3 V
Period of the voltage	T_{sw}	9.7 μs
Switching time period	τ	2.47 μs
Rising time	t_r	70 ns
Falling time	t_f	29 ns
Duration of the turn-off ringing	t_{on}	2.5 μs
Amplitude of the turn-off ringing	A_{r1}	41.5 V
Period of the turn-off ringing	T_{r1}	42.3 ns

According to the mathematical definition [14], the formula for calculating the Fourier series corresponding to trapezoid could be expressed in Equation (2).

$$C_{n\_t} = -j\frac{A_{SW}}{2\pi n}e^{-\frac{jn\pi(\tau+t_r)}{T_{SW}}}\left[\sin c\left(n\frac{t_r}{T_{SW}}\right)e^{-\frac{jn\pi\tau}{T_{SW}}} - \sin c\left(n\frac{t_f}{T_{SW}}\right)e^{-\frac{jn\pi\tau}{T_{SW}}}\right] \tag{2}$$

In order to approximate the experimental waveform, the difference of turn-on time and turn-off time was also taken into consideration.

Before calculating Fourier series of the ringing waveform, it is better to clarify the corresponding time-domain expressions. The ringing shown in Figure 4 begins after the switch turns off (t_1), and will last until the switch turns on (t_2). Therefore, it could be written as a piecewise function, as is presented in Equation (3):

$$V_r(t) = \begin{cases} 0, & 0 < t < t_1 \\ \frac{A_{r1}\sin\omega_{r1}(t-t_r)}{e^{k_{r1}(t-t_r)}}, & t_1 \le t < t_2 \\ 0, & t_2 \le t < T \end{cases} \tag{3}$$

According to the definition, the formula for calculating Fourier series can be written as in Equation (4). Then, the Fourier series of the ringing part can be calculated by substituting Equation (3) into Equation (4).

$$C_{n\_r} = \frac{1}{T}\int_{\langle T\rangle} V_r(t)\cdot e^{-jn\omega}\cdot dt \tag{4}$$

Finally, the Fourier coefficient of the overall switch waveform would be expressed as in Equation (5).

$$C_n = C_{n\_t} + C_{n\_r} \tag{5}$$

Figure 5 shows the experimental setup for capturing the voltage waveform and measuring the conducted EMI noise. The connection order is the same with configuration in Figure 1. In addition, the spectrum analyzer is connected with LISN to receive and analyze the signal, and the PC is used to display and record the spectra results. The EUT in this article is the conventional boost circuit and the different balanced boost circuits.

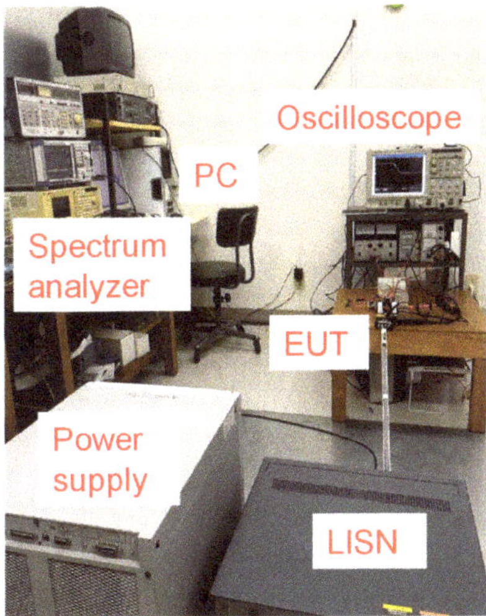

Figure 5. The experimental setup of conducted noise measurement.

According to Equations (2)–(5) and the measured data, the Fourier series calculation was completed with Matlab 2015b. Figure 6 shows the noise source spectra comparison between experimental and calculation results. The calculation results are plotted in the blue line, and the red experimental results refer to Fast Fourier Transformation (FFT) computed by the oscilloscope. The comparison shows that the calculation results match the experimental result in all frequency ranges. The high-frequency peak is caused by the ringing, and this frequency (around 23.5 MHz) agrees with the frequency of ringing waveform in Figure 3 (T_{r1} = 42.3 ns). The effectiveness of this frequency domain modeling method and the obtained parameters was well verified.

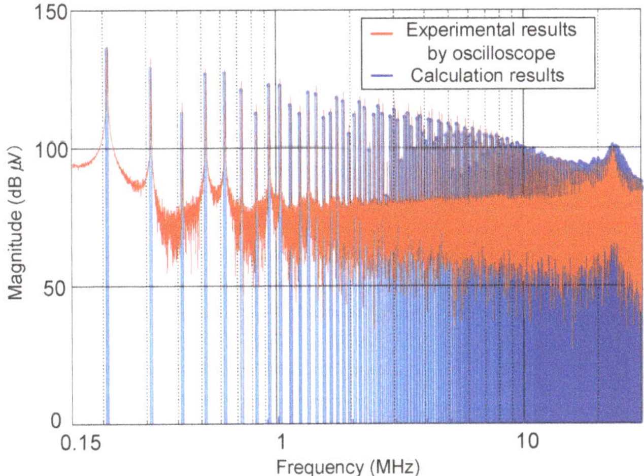

Figure 6. The frequency spectra of switch voltage (noise source) comparison between calculation and measured results.

Furthermore, based on Equation (1), Figure 7 shows the prediction result of CBC by utilizing the calculated noise source. The experimental result measured by LISN is also given in green. It can be observed that the prediction result matches the measurement in the whole frequency band, and this model is able to predict CM noise with good accuracy.

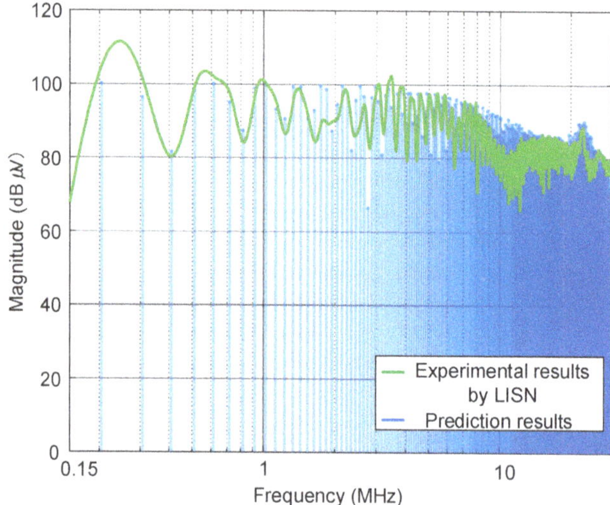

Figure 7. CM noise spectra comparison of a conventional boost converter (CBC) between prediction and measured results.

3. Analysis and Modeling of Impedance Balancing Technique

Instead of adding conventional passive or active EMI filters, the impedance balancing technique could reduce the conducted noise by changing the topology of CBC. The configuration of balanced boost converter along with LISN is shown in Figure 8. Compared with Figure 1, the power inductor L was split into two parts (L_1 and L_2) and another capacitor C_2 was placed between the N phase and the ground. This action changes the path of the noise current without affecting the efficiency of the circuit.

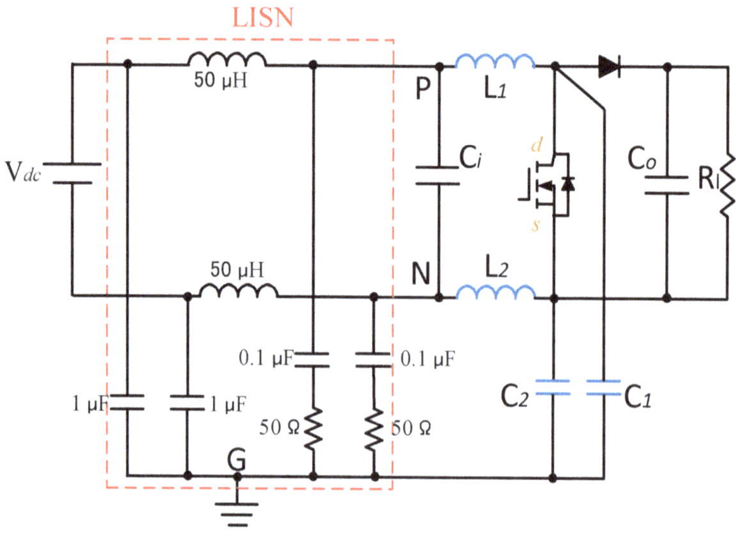

Figure 8. Configuration of the balanced boost converter and LISN.

3.1. CM Noise Model of Impedance Balanced Boost Circuit

The corresponding CM noise model of balanced boost circuit is shown in Figure 9a. The model includes the equivalent parallel capacitor (EPC) and equivalent parallel resistor (EPR) of each inductor, as well asthe equivalent series inductor (ESL) and equivalent series resistor (ESR) of each capacitor. These parameters may affect the self-resonant frequency and cause impedance unbalancing as well. In addition, a more simplified model obtained according to the Thevenin theorem is shown in Figure 9b.

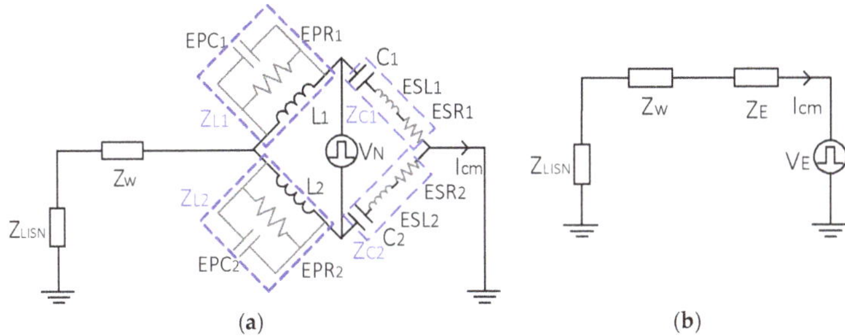

Figure 9. (**a**) CM noise model of the impedance balanced boost circuit; (**b**) simplified CM noise model by the Thevenin theorem.

V_E and Z_E in the figure are the equivalent noise source and the equivalent impedance, respectively. V_N is the same with CBC since the driver circuit of the switch is the same with CBC, and the implement of balancing impedance would not affect the efficiency of the circuit. The relationships between V_E, Z_E, V_N, and the impedance bridge arms are shown in Equations (6) and (7).

$$Z_E = \frac{Z_{L_1} Z_{L_2}}{Z_{L_1} + Z_{L_2}} + \frac{Z_{C_1} Z_{C_2}}{Z_{C_1} + Z_{C_2}} \tag{6}$$

$$V_E = \left(\frac{Z_{L_2}}{Z_{L_1} + Z_{L_2}} - \frac{Z_{C_2}}{Z_{C_1} + Z_{C_2}} \right) V_N \tag{7}$$

where Z_{L1}, Z_{L2}, Z_{C1} and Z_{C2} are the impedance of each arm. Furthermore, the voltage of LISN can be expressed as in Equation (8), which has a similar form as Equation (1).

$$V_{LISN} = \frac{Z_{LISN}}{Z_{LISN} + Z_w + Z_E} V_E \tag{8}$$

Specifically, no CM current would flow through LISN if the impedance of each bridge arm could satisfy the condition in Equation (9):

$$\frac{Z_{L_1}}{Z_{L_2}} = \frac{Z_{C_1}}{Z_{C_2}} = n \tag{9}$$

where n represents the impedance ratio.

According to the above analysis, the most important step of a balanced circuit is selecting the appropriate impedance arms and keeping the impedance ratio (n) stable during all the target frequency bands. The coupled inductor has been previously used in publications due to its smaller size [19–21]. However, unexpected problems have also arisen. Firstly, the coupling coefficient not only affects the inductance value after decoupling but also imposes specific requirements on the selection of core. In [19], the toroidal core has been replaced because it is hard to acquire a high coupling coefficient when the turn ratio is high. Secondly, the parasitic capacitors between the primary side and secondary side make the route of CM noise propagation really complicated. This component has greatly increased the difficulty of modeling analysis and may cause poor reduction results at the same time. Therefore, two isolated inductances were adopted in this paper to prevent these consequences.

In this article, as the power inductor in the CBC is 500 μH, two 250 μH inductors (L_1, L_2) and two 1nF Y-capacitors (C_1, C_2) were utilized to build the balanced circuit. The magnitude spectra of impedance bridge arms are shown in Figure 10. The figure shows that both inductors and capacitances are well matched over the frequency range. The condition of balancing impedance is properly satisfied.

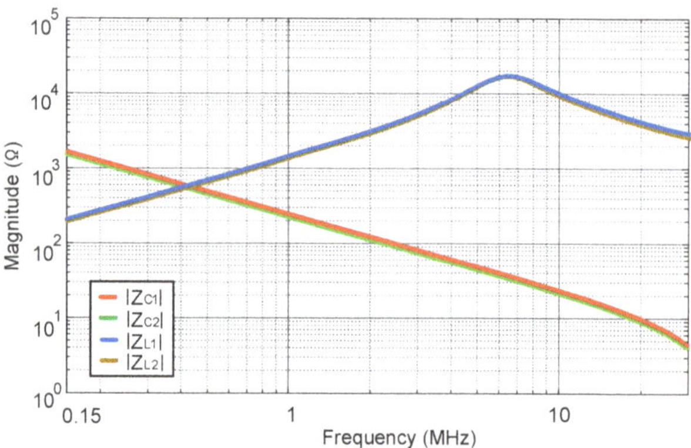

Figure 10. Magnitude comparison of bridge arms impedance (Z_{C1} VS. Z_{C2} and Z_{L1} VS. Z_{L2}).

Figure 11 shows the experimental CM noise spectrum of balanced boost circuit, which is plotted in red. Meanwhile, the green envelope is the CM noise of CBC (same as in Figure 7). By utilizing the balanced technique, the noise level was well attenuated from 0.5 MHz and upwards. The effect of this technique is obvious, as much as 40 dB reduction was done at some frequency points. Nevertheless, the noise peak at low frequency bands degraded the performance. The relative analysis and solution are given in the following sections.

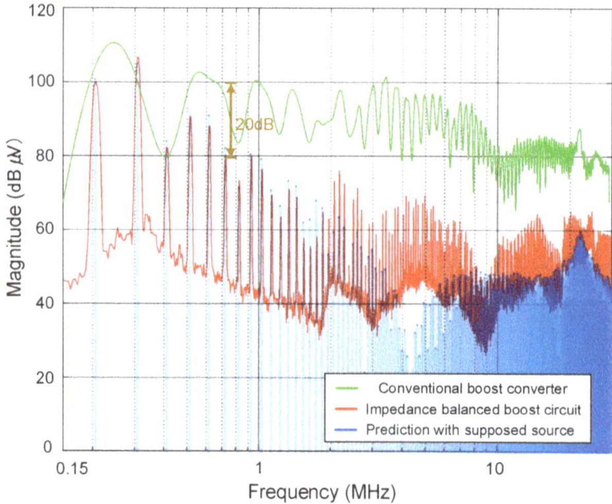

Figure 11. CM noise spectra comparison of balanced impedance circuit.

3.2. Noise Source Evaluation and Noise Spectrum Prediction

As is shown in Figure 11, the CM noise of balanced boost circuit did not dropped to zero, even if the impedance arms are almost balanced. This means that Equation (7) could not calculate the equivalent noise source accurately. This error may be caused by the interference of the PCB layout and the inevitable mismatch of the impedance. Therefore, this article proposed a novel method to evaluate the attenuation level of the noise source and give prediction accordingly.

Comparing the voltage of LISN in the CBC and the balanced boost circuits, which are represented in Equations (1) and (8), respectively, it should be noted that the two formulas have the same form and the difference is the impedance Z_E and Z_{Cg} in the denominator. This also means that the first half of Equations (1) and (8) are the same when Z_E equals to Z_{Cg}, and then the differences of the two noise sources can be directly deducted from the CM noise spectra, as shown in Figure 11.

The equivalent impedance of balanced boost circuit Z_E could be calculated by Equations (6), or measured directly. The error between the two results is small. Referring to Equation (6), Z_{E0} represents the entire impedance of 250 μH inductors (L_1, L_2) paralleled with 1 nF capacitors (C_1, C_2). Figure 12 shows the impedance variation of Z_{Cg} and Z_{E0} through frequency. The Z_{Cg} is larger than Z_{E0} from 0.15 MHz to 0.69 MHz, and then the Z_{E0} becomes larger from 0.69 MHz.

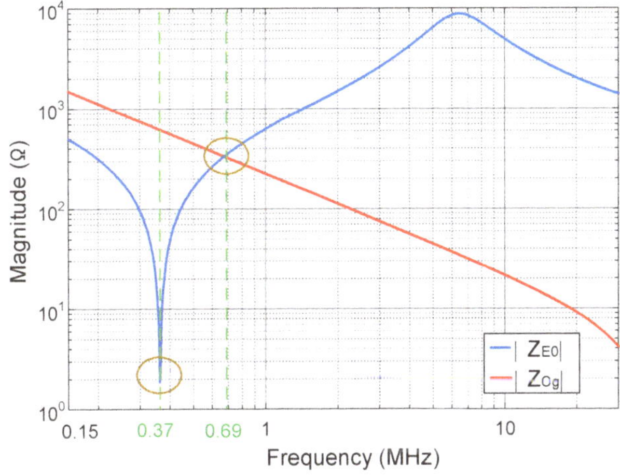

Figure 12. Magnitude spectra comparison between Z_{E0} (250 μH and 1 nF) and Z_{Cg}.

According to Figure 12, it should be noted that Z_{Cg} equals Z_{E0} at the point of 0.69 MHz, where the noise of balanced circuit is 80 dBμV and the CBC is about 100 dBμV. This spectra difference equals to 20 dBμV, which means that V_N is about 10 times of V_E according to the definition. Therefore, this paper assumed that V_E is one-tenth of V_N and then substituted other measured impedance values into Equations (8) for calculation.

The prediction results are plotted in translucent blue in Figure 11 and are basically consistent with the measured spectrum except for the medium frequency band. This comparison confirms the hypothesis.

In addition, the resonance frequency of impedance arms could be calculated by Equations (10).

$$f_c = \frac{1}{2\pi\sqrt{(L_1//L_2)(C_1//C_2)}} \tag{10}$$

Specially, when L_1 and L_2 equal to 250 μH, C_1 and C_2 equal to 1nF, the series resonance of Z_{E0} happens at 0.37 MHz, which agrees with the noise peak in Figure 10. Therefore, it is necessary to find corresponding countermeasures to optimize the known impedance balancing technique.

4. Optimized Schemes with Redesigned Impedance Arms

As shown in Figure 11, the noise peak occurs at the resonant frequency; thus, several researchers have proposed different methods to solve the spike problem. In [17], the authors changed the inductance value and added a pF-class capacitor to cooperate with the impedance ratio in Equation (9). However, this may push the noise spike to the middle or high frequency band. The designing method of the inductor in [21] achieved satisfactory results, but this result is based on an additional CM choke filter. As opposed to previous methods, this paper adjusted the values of passive elements to move the resonance peak out of the conductive frequency range.

As was mentioned in the previous section, a low-frequency peak occurs because of the series resonance among the four impedance bridge arms. On the other hand, it is hard to further suppress the equivalent noise source because the impedance arms are well matched, especially at a low frequency. Thus, the more practical method is to redesign the value of inductors or capacitors. Without changing the balancing configuration shown in Figure 8. This paper has proposed two alternative schemes to attenuate the noise peak according to Equations (10). One is increasing the inductor to 1.4 mH and the other is increasing the value of Y-capacitors to 4.7 nF. The following sections will investigate and compare the two approaches.

4.1. Balanced Boost Circuit with 1.4 mH Inductors and 1 nF Y-capacitors

After calculation using Equations (10), the inductance should be larger than 1.3 mH if the capacitors C_1 and C_2 are kept constant. Here, two1.4mH inductors were utilized to replace the 250 μH inductors. The experimental results were recorded and drawn with black in Figure 13, compared with the former balanced boost circuit in red and the green envelope of CBC. These experimental results show that the proposed balanced scheme with 1.4 mH inductors and 1nF capacitors could attenuate noise level over all frequency range compared with CBC. Furthermore, the spike at 0.37 MHz was effectively suppressed. However, the noise at high frequency increased in contrast with the former balanced boost circuit (250 μH and 1 nF). The valley at 0.8 MHz is due to the self-resonance between the 1.4 mH inductor and its EPCs.

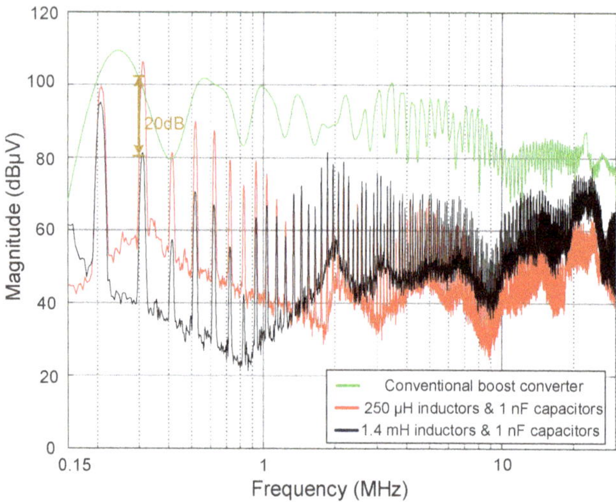

Figure 13. CM noise comparison after replacing the 250 μH inductors with 1.4 mH inductors.

Referring to Equations (6), Z_{E1} identifies the whole impedance including 1.4 mH inductors and 1nF Y-capacitors. Figure 14 gives its magnitude in black and makes comparison with Z_{Cg} in red. Note that Z_{E1} equals to Z_{Cg} at 0.24 MHz, which means that the noise source attenuation degree could be speculated quickly With the same analysis method as the one in Section 3. According to Figure 13, the noise spectra difference around 0.24 MHz reached about 20 dBμV, which indicates that the equivalent noise source V_E in the optimized scheme was attenuated well compared with CBC. However, a drawback of this approach is the increased weight and volume of the circuit.

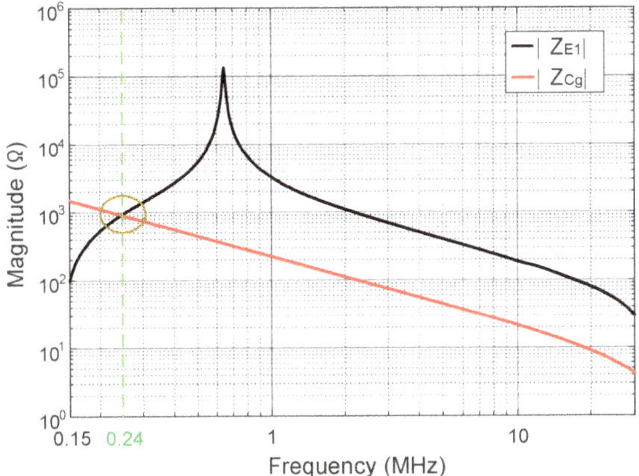

Figure 14. Impedance comparison between Z_{E1} (1.4 mH and 1 nF) and Z_{Cg}.

4.2. Balanced Boost Circuit with 250 μH Inductors and 4.7 nF Y-capacitors

Other than using a pair of larger inductors, this scheme utilized greater Y-capacitors to decrease the series resonant frequency. After calculation, 4.7 nF capacitors were selected to replace the former 1nF capacitors and the other experimental conditions were kept the same.

The CM noise spectra comparison is shown in Figure 15. The purple curve is for the case of an optimized solution after replacing the 1 nF Y-capacitors with 4.7 nF, the green one is the noise envelope of CBC, and the red one is for the former balanced circuit boost (250 μH and 1 nF). The comparison shows that after utilizing the scheme with 4.7 nF capacitors, a low-frequency spike was also moved out from the frequency band of interest and the CM noise was significantly attenuated. Meanwhile, at a high frequency, the noise after adopting 4.7 nF capacitors is lower than the former case using 1 nF capacitors.

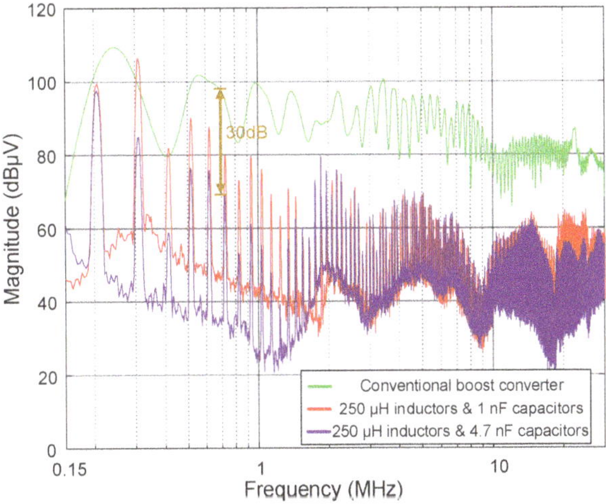

Figure 15. CM noise comparison after replacing the 1 nF Y-capacitors with 4.7 nF Y-capacitors.

Similarly to Z_{E1}, Z_{E2} represents the whole bridge impedance of 250 µH inductors and 4.7 nF Y-capacitors. The comparison between Z_{E2} (purple) and Z_{Cg} (red) is shown in Figure 16. The resonant frequency of optimized impedance Z_{E2} was removed to around 0.16 MHz and the performance at high frequency is almost the same compared to that in Figure 12. At 0.68 MHz, where Z_{E2} equals Z_{Cg}, about 30 dB reduction was achieved according to Figure 15. Compared with the former balanced boost circuit (250 µH and 1 nF), the noise source was significantly attenuated and the performance was improved over the entire frequency range.

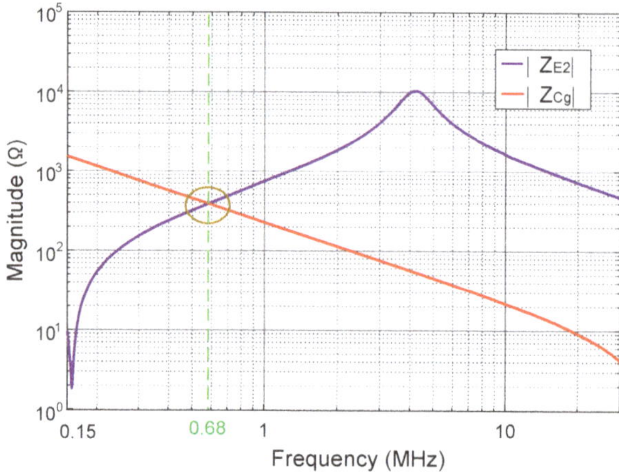

Figure 16. Impedance comparison between Z_{E2} (250 µH and 4.7 nF) and Z_{Cg}.

There may be some concerns that the greater capacitor may increase the CM current and exceed relative standards. Figure 17 shows the CM currents comparison between CBC and the optimized balanced boost circuit (250 µH and 4.7 nF). The results show that with optimized schemes, the CM current becomes even smaller because the noise source has been effectively attenuated.

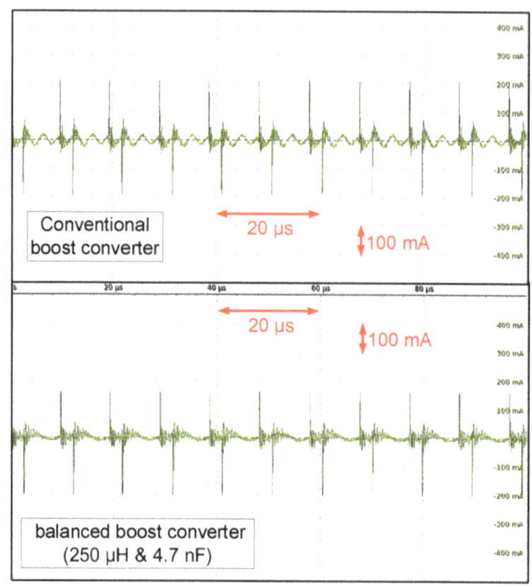

Figure 17. Experimental CM current waveforms comparison between CBC and optimized impedance balancing boost converter (250 µH and 4.7 nF).

In summary, the optimized scheme of changing the capacitors is more effective than replacing the inductors, which not only considers the attenuation results but also the cost and the difficulty of implementation.

A comparative investigation between the proposed work and the previously published relevant arts was carried out and is addressed in Table 3. The comparison reveals the effectiveness and practicability of the proposed schemes.

Table 3. Comparison of noise optimization effects.

	Noise Reduction Effect	Attenuated the Resonance Peak?	Required Additional Filter?	Practicability and Feasibility
[16]	Medium	No	Yes	Medium
[17]	Medium	No	No	Medium
[18]	Limited	No	Yes	Limited
[19,20]	Medium	Yes	Yes	Medium
Proposed work	Medium	Yes	No	High

5. Conclusions

Focusing on the noise attenuation with impedance balancing technique, a frequency domain modeling method and two optimized schemes are introduced in this paper. The accurate noise source model was extracted by capturing the switch voltage waveform and then calculating its Fourier series. Furthermore, for both noise source and CM noise of the conventional boost circuit, the accuracy of the noise model were verified by comparing the prediction results with the experimental measurements.

Then, the impedance balancing technique was adopted to attenuate the noise. Although this technique is effective to some extent, it suffers from noise spikes and also lacks a model that can predict the noise quantitatively. In this research, the noise source attenuation degree was evaluated by combining the noise spectra and impedance information. Furthermore, a quantitative prediction based on the proposed noise model is provided. A comparison between prediction and experimental results was conducted and the accuracy of the proposed noise model was practically verified.

Based on the basic impedance balancing configuration and the calculation formula of resonance frequency, two optimized schemes were proposed to deal with the noise peak in the low-frequency band. The size of the impedance arms was redesigned and the two schemes were investigated in different aspects. Finally, the experimental results were presented to demonstrate the analysis. The contribution of this research is enhancing the feasibility and effectiveness of the known impedance balancing technique.

Author Contributions: Conceptualization, methodology, investigation, S.Z., B.Z., M.S., and G.M.D.; validation, S.Z. and Q.L.; writing—review and editing S.Z. and G.M.D.; supervision M.S. and G.M.D.; project administration, E.T. All authors have read and agreed to the published version of the manuscript.

Funding: This research received no external funding.

Conflicts of Interest: The authors declare no conflict of interest.

References

1. Mainali, K.; Oruganti, R. Conducted EMI Mitigation Techniques for Switch-Mode Power Converters: A Survey. *IEEE Trans. Power Electron.* **2010**, *25*, 2344–2356. [CrossRef]
2. Tarateeraseth, V.; Hu, B.; See, K.Y.; Canavero, F.G. Accurate Extraction of Noise Source Impedance of an SMPS Under Operating Conditions. *IEEE Trans. Power Electron.* **2010**, *25*, 111–117. [CrossRef]

3. Yang, L.; Lu, B.; Dong, W.; Lu, Z.; Xu, M.; Lee, F.C.; Odendaal, W.G. Modeling and characterization of a 1 KW CCM PFC converter for conducted EMI prediction. In Proceedings of the Nineteenth Annual IEEE Applied Power Electronics Conference and Exposition, APEC'04, Anaheim, CA, USA, 22–26 February 2004; Volume 2, pp. 763–769.
4. Lai, J.; Huang, X.; Pepa, E.; Chen, S.; Nehl, T.W. Inverter EMI modeling and simulation methodologies. *IEEE Trans. Ind. Electron.* **2006**, *53*, 736–744.
5. Gubia, E.; Sanchis, P.; Ursua, A.; Lopez, J.; Marroyo, L. Frequency domain model of conducted EMI in electrical drives. *IEEE Power Electron. Lett.* **2005**, *3*, 45–49. [CrossRef]
6. Pei, X.; Zhang, K.; Kang, Y.; Chen, J. Prediction of common mode conducted EMI in single phase PWM inverter. In Proceedings of the 2004 IEEE 35th Annual Power Electronics Specialists Conference (IEEE Cat. No.04CH37551), Aachen, Germany, 20–25 June 2004; Volume 4, pp. 3060–3065.
7. Liu, Q.; Wang, F.; Boroyevich, D. Modular-Terminal-Behavioral (MTB) Model for Characterizing Switching Module Conducted EMI Generation in Converter Systems. *IEEE Trans. Power Electron.* **2006**, *21*, 1804–1814.
8. Bishnoi, H.; Mattavelli, P.; Burgos, R.; Boroyevich, D. EMI Behavioral Models of DC-Fed Three-Phase Motor Drive Systems. *IEEE Trans. Power Electron.* **2014**, *29*, 4633–4645. [CrossRef]
9. Kerrouche, B.; Bensetti, M.; Zaoui, A. New EMI Model with the Same Input Impedances as Converter. *IEEE Trans. Electromagn. Compat.* **2019**, *61*, 1072–1081. [CrossRef]
10. Mainali, K.; Oruganti, R. Simple Analytical Models to Predict Conducted EMI Noise in a Power Electronic Converter. In Proceedings of the IECON 2007—33rd Annual Conference of the IEEE Industrial Electronics Society, Taipei, Taiwan, 5–8 November 2007; pp. 1930–1936.
11. Jin, M.; Weiming, M. Power Converter EMI Analysis Including IGBT Nonlinear Switching Transient Model. *IEEE Trans. Ind. Electron.* **2006**, *53*, 1577–1583. [CrossRef]
12. Xia, Y.; Gu, Y.; Shen, J. Conducted EMC Modeling for EV Drives Considering Switching Dynamics and Frequency Dispersion. In Proceedings of the 2018 IEEE Energy Conversion Congress and Exposition (ECCE), Portland, OR, USA, 23–27 September 2018; pp. 4195–4202.
13. Xiang, Y.; Pei, X.; Zhou, W.; Kang, Y.; Wang, H. A Fast and Precise Method for Modeling EMI Source in Two-Level Three-Phase Converter. *IEEE Trans. Power Electron.* **2019**, *34*, 10650–10664. [CrossRef]
14. Zhang, B.; Zhang, S.; Li, H.; Shoyama, M.; Takegami, E. An Improved Simple EMI Modeling Method for Conducted Common Mode Noise Prediction in DC-DC Buck Converter. In Proceedings of the IEEE ICDCM'19, Matsue, Japan, 20–23 May 2019; pp. 1–6.
15. Kim, T.; Feng, D.; Jang, M.; Agelidis, V.G. Common Mode Noise Analysis for Cascaded Boost Converter with Silicon Carbide Devices. *IEEE Trans. Power Electron.* **2017**, *32*, 1917–1926. [CrossRef]
16. Shoyama, M.; Li, G.; Ninomiya, T. Balanced switching converter to reduce common-mode conducted noise. *IEEE Trans. Ind. Electron.* **2003**, *50*, 1095–1099. [CrossRef]
17. Wang, S.; Kong, P.; Lee, F.C. Common Mode Noise Reduction for Boost Converters Using General Balance Technique. *IEEE Trans. Power Electron.* **2007**, *22*, 1410–1416. [CrossRef]
18. Xing, L.; Sun, J. Conducted Common-Mode EMI Reduction by Impedance Balancing. *IEEE Trans. Power Electron.* **2012**, *27*, 1084–1089. [CrossRef]
19. Kong, P.; Wang, S.; Lee, F.C. Improving balance technique for high frequency common mode noise reduction in boost PFC converters. In Proceedings of the 2008 IEEE Power Electronics Specialists Conference, Rhodes, Greece, 15–19 June 2008; pp. 2941–2947.
20. Kong, P.; Wang, S.; Lee, F.C. Common Mode EMI Noise Suppression for Bridgeless PFC Converters. *IEEE Trans. Power Electron.* **2008**, *23*, 291–297. [CrossRef]
21. Zhang, Y.; Zheng, F.; Wang, Y. Alleviation of Electromagnetic Interference Noise Using a Resonant Shunt for Balanced Converters. *IEEE Trans. Power Electron.* **2015**, *30*, 4762–4773. [CrossRef]

Article

Electromagnetic Susceptibility of Battery Management Systems' ICs for Electric Vehicles: Experimental Study

Orazio Aiello

Department of Electronics and Telecommunications, Politecnico di Torino, Corso Duca degli Abruzzi 24, I-10129 Torino, Italy; orazio.aiello@polito.it

Received: 20 January 2020; Accepted: 13 March 2020; Published: 19 March 2020

Abstract: The paper deals with the susceptibility to electromagnetic interference (EMI) of battery management systems (BMSs) for Li-ion and lithium-polymer (LiPo) battery packs employed in emerging electric and hybrid electric vehicles. A specific test board was developed to experimentally assess the EMI susceptibility of a BMS front-end integrated circuit by direct power injection (DPI) and radiated susceptibility measurements in an anechoic chamber. Experimental results are discussed in reference to the different setup, highlighting the related EMI-induced failure mechanisms observed during the tests.

Keywords: battery management system (BMS); Li-ion battery pack; electric vehicles (EVs); hybrid electric vehicles (HEVs); IC-level EMC; susceptibility to electromagnetic interference (EMI); direct power injection (DPI); anechoic chamber

---

1. Introduction

A greater and greater increase in the amount of electronic devices is expected in new vehicles to make cars capable of running self-diagnostics and interacting with the surrounding environment. On-board electronics systems for implementing safety features to reduce the number of accidents and fatalities on the roads have to properly operate in any operating conditions [1]. Smart power integrated circuits (ICs) employed in automotive front-end electronics have stringent requirements in terms of accuracy, even in an electromagnetically polluted environment, where relevant conducted and radiated interference are generated by the electric powertrain, by on-vehicle electronics and by mobile phones and/or other information and communication equipment carried by the driver and by the passengers in the cockpit [1]. Therefore, the susceptibility to electromagnetic interference (EMI) of smart power electronics and its on-board monitoring and control functions (i.e., thermal shutdown, current sensors and overvoltage protection) has to be considered. This to prevent malfunctions and guarantee the correct/expected functioning of the electronic system in any operating conditions [1–6].

Critical safety EMI-induced failures could be also a major threat to the safety in emerging electric and hybrid electric (EV/HEV) vehicles powered by batteries [7]. A battery management system (BMS) IC manages the state of charge of the battery pack, protecting it from operating outside its safe operating conditions [8–13]. Electromagnetic interference can be easily picked up by the long wires that connect BMS front-end ICs to each other, to the BMS control unit, to the terminals of the electrochemical cells and to the temperature sensors, which are spatially distributed over the whole battery pack module [14], and can easily impair the operation of data acquisition circuits [15–18].

Battery packs based on the most advanced lithium-ion (Li-ion) and lithium-polymer (LiPo) electrochemical technologies are nowadays the only viable options to address the challenging demands in terms of the electric energy storage and deliverable power per unit mass of electric vehicles (EVs) and hybrid electric vehicles (HEVs) [19–22]. Unfortunately, unlike lead-acid batteries and other more conventional electrochemical accumulators, Li-ion and LiPo cells can be permanently damaged and can also originate life-threatening hazards such as fires and explosions in the event of overdischarging, overcharging and/or overtemperature operation [21–23]. An electronic battery management system (BMS), which quickly detects the onset of dangerous conditions and takes the appropriate countermeasures to avoid hazards, is therefore necessary to safely operate Li-ion and LiPo cells in vehicles [21–23]. A BMS, which is schematically depicted in Figure 1, typically includes several front-end modules that acquire critical cell information, such as terminal voltages and temperatures, and a digital control unit that runs specific control and management algorithms.

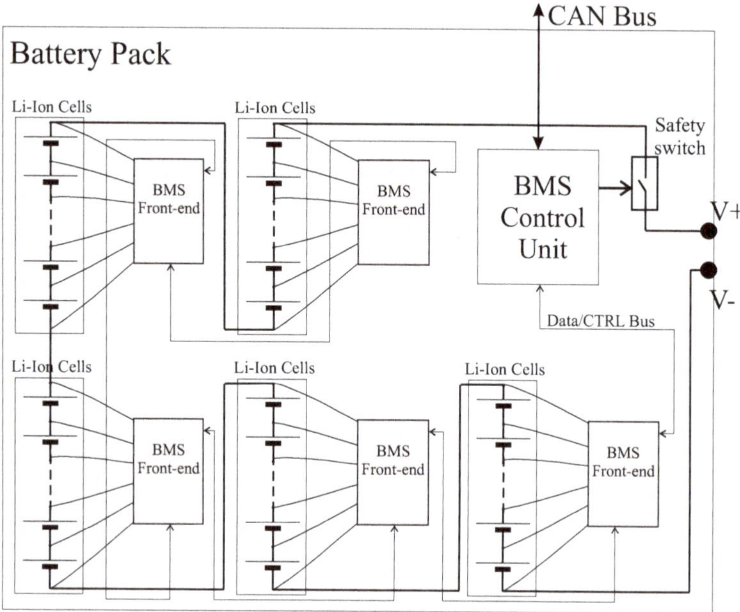

Figure 1. Architecture of a battery management system (BMS) for EV/HEV applications.

The typical BMS application scenario in Figure 1 and the EMI susceptibility issues of BMS systems for electric vehicles are addressed in this paper on the basis of the results of direct power injection (DPI) tests (IEC-62132-4 standard [24,25]) and radiated susceptibility tests (ISO11452-2 standard [26]) on a commercial smart power IC, widely employed as a BMS front-end in EV/HEV applications. In particular, the specific susceptibility level of the main IC pins and their different EMI-induced failure mechanisms were observed. The effectiveness of filtering techniques that can be adopted to enhance the immunity to EMI of a BMS, is also discussed.

The paper is organized as follows: in Section 2, the BMS IC is presented, and the printed circuit board (PCB) developed to perform EMC tests is introduced; in Section 3 the test bench and the test procedure adopted for DPI measurements are described. DPI test results are presented in Section 4 and discussed in Section 5. The results of radiated susceptibility measurements are presented in Section 6 and compared to the DPI tests in Section 7, and finally, in Section 8, some concluding remarks are drawn.

2. The BMS Front-End for the Tests

In order to investigate the susceptibility to EMI of a BMS for electric vehicles and perform the related conducted (DPI) and radiated (in anechoic chamber) tests, a PCB was specifically designed, and it is described in this Section.

2.1. BMS Front-End IC Structure and Operation

The simplified block diagram of the BMS front-end IC, which is considered in what follows as the device under test (DUT), is reported in the pink box of Figure 2. Such an IC is designed to monitor the terminal voltages of up to twelve series-connected electrochemical cells using a 12-bit analog-to-digital converter (ADC). To that end, the IC includes 12 cell input pins, internally connected to the ADC's differential input by a 12-channel, isolated, high-voltage analog multiplexer to measure the (differential) voltages across each of twelve series-connected cells, whose common-mode voltage can be up to 50 V with respect to the IC reference.

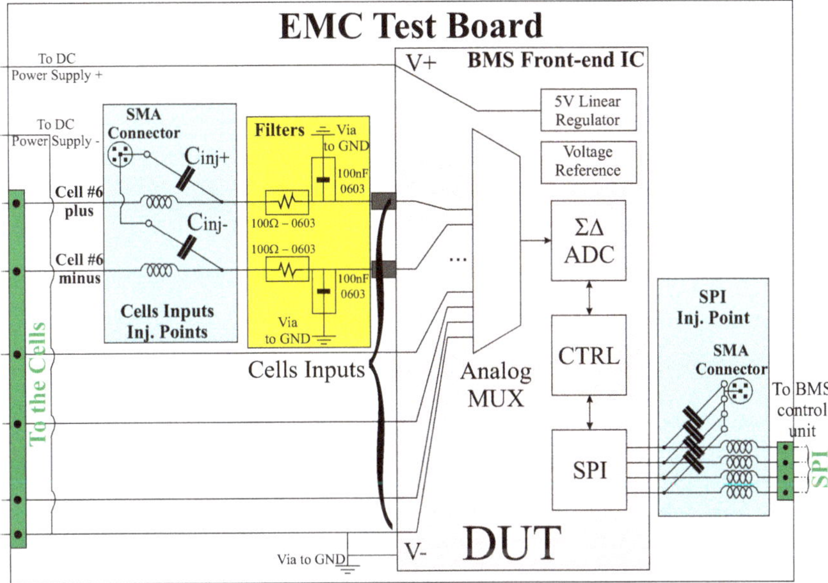

Figure 2. Simplified schematic diagram of the BMS integrated circuit (IC) direct power injection (DPI) test board.

The same IC can be also exploited for cell temperature measurements by using external negative temperature coefficient (NTC) sensors and including power drivers suitable for performing passive Li-ion cell voltage equalization. These functions have not been considered the susceptibility assessment and will not be mentioned hereafter.

The IC is designed to be operated by an external BMS control unit via a serial peripheral interface (SPI), through which acquired data can also be retrieved. The same SPI interface can be employed to connect the BMS IC to N similar devices in a daisy chain structure, as shown in Figure 1, so as to monitor up to $12N$ cells, addressing the requirements of high voltage battery packs including 100 or more series-connected cells. The frame of the specific SPI protocol implemented in the DUT includes a packet error code (PEC), by which SPI bus errors can be detected on the basis of the received data.

The IC operates from a DC supply voltage ranging from 10 V up to 55 V, which can be obtained either from the electrochemical cells to be monitored or from an external isolated DC/DC converter, and includes an internal voltage reference for the ADC and a 5 V linear voltage regulator to supply the low voltage analog and digital circuitry.

2.2. DPI Test Board

Firstly, the susceptibility to EMI of the DUT has been tested by the DPI method [24,25]. The dual layer PCB reported in Figure 2 has been designed to inject RF power into the IC pins which are connected to the external wires. These possibly long wires are likely to collect a relevant amount of EMI in a realistic EV/HEV application. Two DPI injection points are established to superimpose an RF power into a couple of cell input pins and into the SPI pins of the IC, as depicted in Figure 2.

The cell input injection point includes an SMA connector and two RF coupling networks (each made up of a 6.8 nF capacitor and a 6 μH inductor), designed so as to inject RF power onto the cell input pins connected to the positive and to the negative terminals of cell #6 in the stack, as depicted in Figure 2. A differential RF injection on a single cell input pin (cell #6 plus in Figure 2) can be performed by connecting only the injection capacitor C^{+}_{inj} to the signal terminal of the SMA connector. A common-mode injection can be performed by connecting both the injection capacitors C^{+}_{inj} and C^{-}_{inj} to the SMA signal line.

Figure 2 shows how the cell input injection points include an SMA connector and two RF coupling networks (each made up of a 1 nF CHECK capacitor and a 6 μH inductor). Two *RC* filters with $R = 100\ \Omega$, $C = 100$ nF implemented using 0603 SMD components (sketched in the yellow area) are mounted on the PCB immediately before the IC pins. The SPI injection point includes an SMA connector and four RF coupling networks (each made up of a 1 nF capacitor and a 6 μH inductor), designed to perform DPI on all the SPI inputs at the same time, or alternatively, on a single SPI line.

3. Direct Power Injection Tests

The failure criteria considered in the DPI tests are stated along with the reasoning in this section.

3.1. DPI Test Bench

The PCB in Figure 2 has been employed to perform the DPI method according the setup in Figure 3. The DUT is operated from an external 20V power supply and the cell inputs of the DUT are connected to six series-connected commercial nickel metal hydride (NiMH) cells mounted on a separate board and tied to the DUT cell inputs by twisted wires. The current flowing through the cells during the DPI test is monitored by an amper meter, as shown in Figure 3.

The SPI lines of the DUT are connected to an automotive microcontroller evaluation board, which acts as the BMS control unit shown in Figure 1. Such a BMS control unit includes a controller area network (CAN) bus interface that is connected to a personal computer (PC) via a Vector CANCaseXL dongle. The microcontroller on the BMS control unit is programmed so as to forward the data content of the CAN packets with a specific identifier (ID) to the DUT SPI, and to send back to the CAN bus, with a different ID, the data retrieved from the DUT SPI during the same transaction. In this way, the DUT can be fully operated and monitored from the PC via the CAN bus. The CAN bus and the measurement instruments are managed by Matlab. The BMS IC is configured to be initialized, start the ADC conversion and read the register.

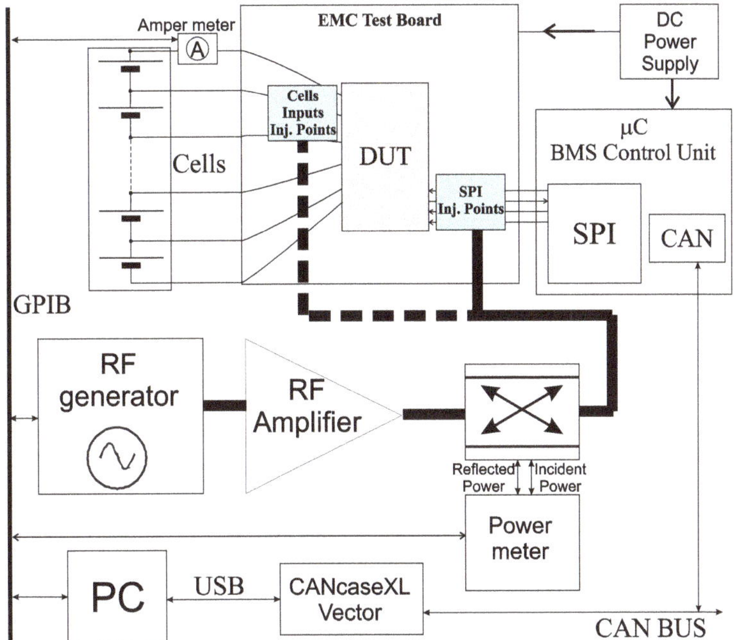

Figure 3. Direct power injection. Experimental test setup.

The continuous wave (CW) RF power injection is performed by means of an RF source connected to a 10 W RF power amplifier with a 1 MHz to 2 GHz bandwidth. The output of the power amplifier is fed to the injection points of the test board through a −20 dB directional coupler, whose forward and reverse coupled ports are connected to an RF power meter so as to monitor the incident and the reflected power. The RF source, the amper meter and the RF power meter are connected to the same PC employed to control the DUT via a general purpose interface bus (GPIB) dongle. Both the CAN bus and the GPIB are fully controlled by the PC in the Matlab environment.

3.2. DPI Test Procedure

During DPI tests, the DUT is operated from the PC in Figure 3 so as to periodically acquire the voltages of all the cells in the battery pack. The acquired data are then retrieved from the DUT SPI by the BMS control unit and forwarded to the PC via the CAN bus, together with the corresponding SPI PEC code. The same operations are first performed without injecting RF power (i.e., with the RF source in Figure 3 off) and then while injecting RF power at a given test frequency. The data retrieved from the DUT, with and without EMI injection, are finally compared and an EMI-induced failure is recorded if one of the following conditions occurs:

1. An SPI transmission error is detected (i.e., the received PEC is not consistent with the received data);
2. An EMI-induced offset in the cell voltages acquired with RF injection exceeding an error threshold V_T is detected.

Taking into account of the accuracy level that is required to safely manage Li-ion cells [23], and the declared maximum error of the IC, an error threshold $V_T = 10$ mV is considered in this paper.

Considering the above failure criteria, DPI tests have been performed for each test frequency in the 1 MHz to 2 GHz bandwidth by increasing the injected RF incident power until a failure is detected. Notice that the IEC 62132-4 frequency bandwidth (1 MHz–1 GHz) has been extended to

2 GHz to include the 1.8 GHz frequency, widely employed in wireless communications. The minimum RF incident power giving rise to the failure is then reported as the DPI immunity level at the test frequency. If no failure is detected at the maximum test incident power $P_{max} = 37$ dBm, no immunity level indication is reported. The results of DPI tests performed according with the above procedure are presented in the next Section.

4. DPI Experimental Results

The DPI immunity tests on the DUT cell inputs and SPI injection points in Figure 2 are reported in this Section. The different failure mechanisms observed are highlighted and the effectiveness of PCB level filtering on cell inputs is discussed.

4.1. Injection of Cell Inputs

The EMI robustness of the DUT undergoing DPI on the cell input injection point is considered comparing a differential (DM) excitation and a common-mode (CM) one. Differential (DM) excitation is performed by connecting only the injection capacitor C^{+}_{inj} to the signal terminal of the SMA connector in Figure 2. Common-mode (CM) excitation is performed by connecting both the injection capacitors C^{+}_{inj} and C^{-}_{inj} to the SMA signal line. Both the tests have been repeated without the RC filters in Figure 2 (i.e., using 0 Ω 0603 SMD resistors and not mounting the filter capacitors) and with the 100 Ω–100 nF *RC* filters prescribed by the IC manufacturer.

The respective measurement results are reported in Figure 4. The DPI immunity level of the DUT without filters is very similar for CM and DM injection and it is mostly in the range of 5–15 dBm over the 1 MHz to 1 GHz bandwidth. A very high susceptibility to EMI (immunity level of less than 0 dBm) can be observed at 16 MHz and harmonic frequencies, which are likely to be related with the internal clock frequency of the sigma-delta converter built-in the BMS IC. As such, EMI-induced failures do not seem to be specifically related to EMI superimposed onto the differential cell voltage to be acquired, but rather to other mechanisms involving the RF voltage between each test pin and the IC reference (ground) voltage.

The presence of *RC* filters provides a significant immunity enhancement in the 20–600 MHz band (a single failure is experienced at 150 MHz at the 37 dBm test level), whereas their effectiveness is lower above 600 MHz. This can be explained by considering that the impedance of the 0603, 100 nF capacitor of the filter, dominated above 20 MHz by the parasitic inductance ESL $\simeq 1$ nH, in series with the PCB track and via inductance ($L_{\text{track}} \simeq 1$ nH and $L_{\text{via}} \simeq 1$ nH), gives rise to a parallel resonance with the input capacitance of the BMS IC ($C_{\text{IN}} \simeq 15$ pF from S-parameters measurements) at a frequency:

$$f_0 = \frac{1}{2\pi\sqrt{(\text{ESL} + L_{\text{track}} + L_{\text{via}})\, C_{\text{IN}}}} \simeq 700\ \text{MHz}. \tag{1}$$

Close to this frequency, the actual impedance of the parallel element of the *RC* filter (capacitor *C* and parasitics) is very high and its filtering effectiveness is therefore impaired.

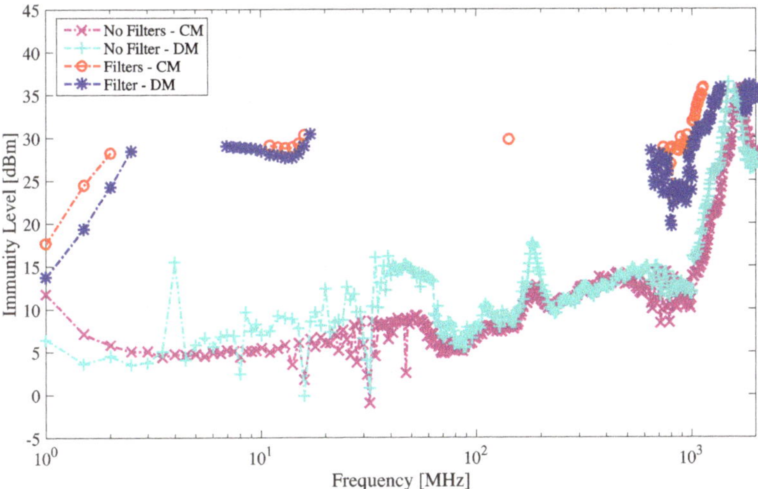

Figure 4. Measured immunity level (expressed in terms of RF incident power) for DPI on the cell input pins: differential (DM) and common-mode (CM) injection, with and without *RC* 100 Ω, 100 nF RC filters in Figure 2.

4.1.1. Test Port Voltage Estimation

To gain further insights into the intrinsic susceptibility level of the IC and the effectiveness of the filters, the peak RF voltage at the test board injection port giving rise to the failures reported in Figure 4 has been estimated on the basis of the reflection coefficient Γ at the test board injection port, measured by a calibrated vector network analyzer (VNA). Taking into account that the peak RF voltage at the test port V_{rf} can be expressed in terms of the RF incident power P_{inc} as

$$V_{rf} = |1 + \Gamma| \sqrt{2 R_G P_{inc}}, \tag{2}$$

where $R_G = 50\ \Omega$ is the reference port resistance, the incident power immunity levels have been translated into the corresponding test port voltages in Figure 5. On this basis, it can noticed that an injected RF voltage with a peak amplitude even lower than 1 V (lower than 0.5 in the worst case) is sufficient to induce a failure in the BMS IC without filters. Moreover, by comparing the failure levels in Figure 4, expressed in terms of incident power, and the results in Figure 5, expressed in terms of injection port voltage, it can be appreciated that the immunity enhancement brought by the filters in the 800 MHz to 2 GHz range is much lower if expressed in terms of the injection port voltage than in terms of incident power, and both the filtered and the unfiltered ICs undergo a failure for an injected EMI peak amplitude of about 3 V.

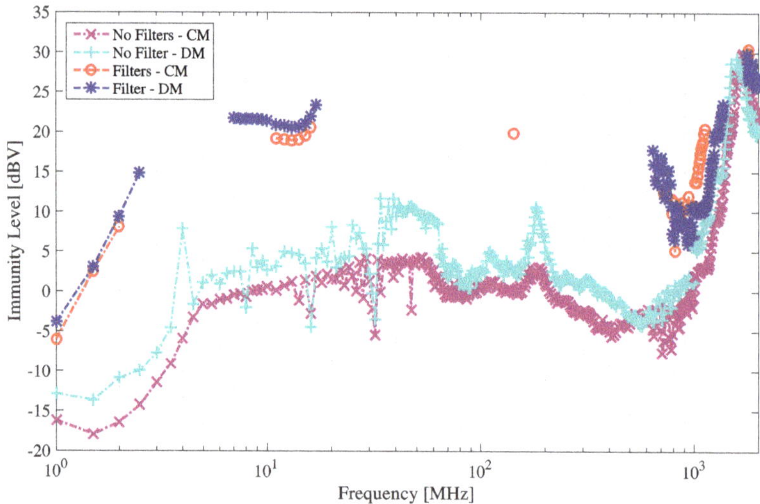

Figure 5. Total RF voltage induced at the DPI injection port corresponding to the DPI immunity levels in Figure 4.

4.1.2. Failure Mechanisms

During EMI induced failures reported in Figure 4, no SPI communication failure (inconsistent PEC) was reported when performing DPI injection on the cell inputs. In all cases, in fact, failures events were related to an EMI-induced offset exceeding 10 mV in the cell voltage readings. Such an offset voltage, defined for each cell *i* as

$$V_{\mathrm{OFF},i} = V_{\mathrm{ADC},i,\mathrm{EMI}} - V_{\mathrm{ADC},i} \tag{3}$$

where $V_{\mathrm{ADC},i,\mathrm{EMI}}$ is the *i*-th cell voltage acquired by the DUT while injecting an RF power corresponding to the immunity level in Figure 4, and $V_{\mathrm{ADC},i}$ is the voltage of the same cell acquired by the DUT without RF power injection, is plotted for each cell in Figure 6 for the unfiltered IC, and in Figure 7 for the IC including RC filters. In both cases, DM injection is considered.

Figure 6 shows how that failures up to about 300 MHz are related to an EMI-induced offset exceeding 10 mV in the acquired voltage of cell #6; i.e., on the cell on which DPI is performed. On the contrary, for EMI frequencies above 300 MHz, all the acquired cell voltages show a similar offset. Such a behavior can be related to the direct propagation of EMI to the internal ADC and/or to its reference voltage source.

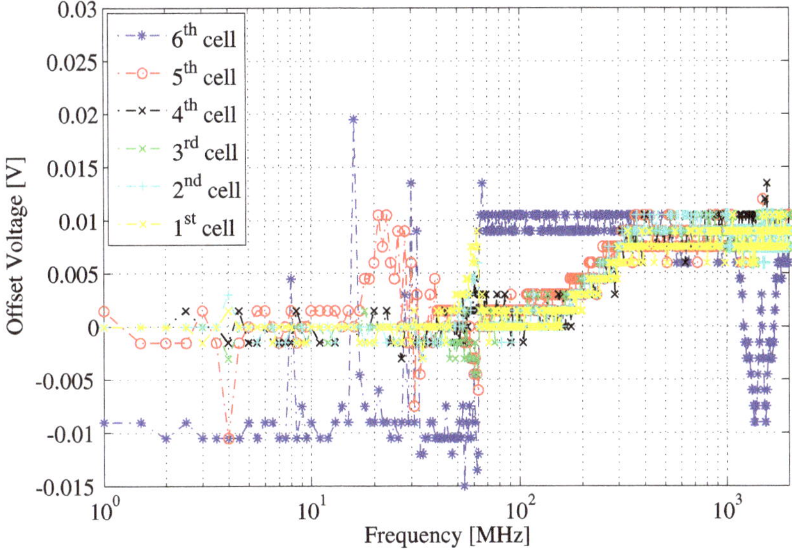

Figure 6. EMI-induced offset in the cell voltage readings obtained in accordance with DPI failure levels of Figure 4 (DM injection performed on the 6th cell positive terminal, without RC filters).

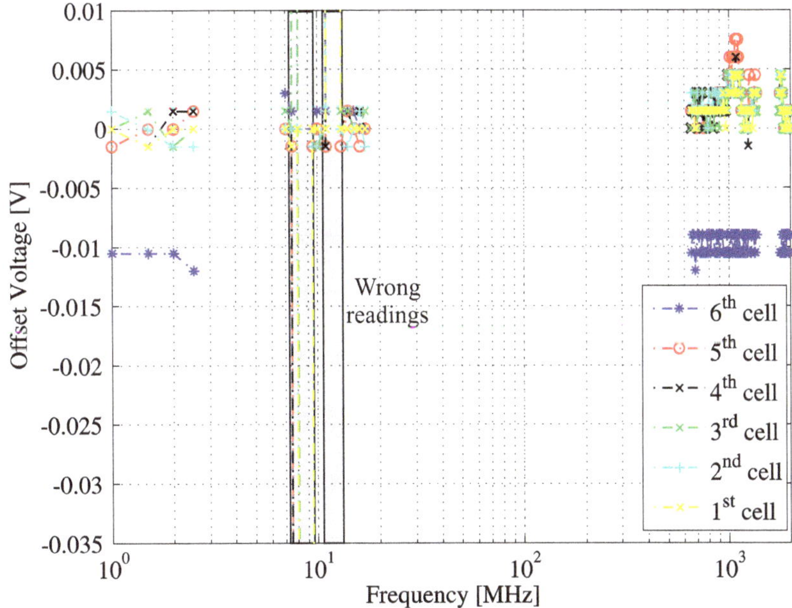

Figure 7. EMI-induced offset in the cell voltage readings obtained in accordance with DPI failures level of Figure 4 (DM injection performed on the 6th cell positive terminal, with RC filters).

Figure 7 shows a different failure mechanism for the filtered device undergoing RF DPI in the 8–12 MHz band (reported as a *wrong reading* gray area). In this case, injected EMI gives rise to completely wrong ADC readings, corresponding to the maximum or to the minimum values of the ADC range.

At lower and higher frequencies the failure mechanisms related to the offset presence are observed. Similar results have been obtained performing CM injection.

Finally, the DC current delivered by the battery pack and measured by the DC amper meter (see Figure 3) in accordance with the DPI failure levels shown in Figure 4, is plotted in Figure 8 for DM DPI performed without (a) and with (b) the RC input filters in Figure 2. Notice that such a current does not include the DC current sunk by the IC for its operation, and no external load is connected to the cells. Figure 8 shows how such a DC current delivered by the battery pack in accordance with the DPI failure levels, (whose measured value without EMI is about 170 μA), can be significative of being affected by the injected disturbances if the RC filters are not included, reaching 6 mA for a 40 MHz EMI injection frequency. It is interesting to observe that the measured DC current is negative (current delivered to the battery pack) for most EMI injection frequencies.

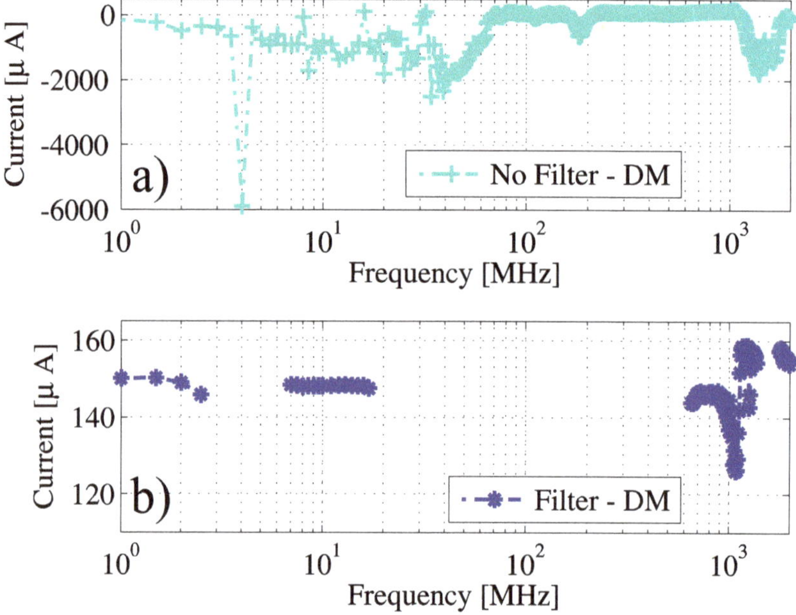

Figure 8. DC current absorbed from the battery pack by the BMS IC cell inputs in accordance with DPI failures levels in Figure 4 for DM injection performed on the 6th, without (**a**) and with (**b**) the RC filters.

4.2. SPI Injection

The immunity level of the DUT undergoing DPI on the SPI injection point in Figure 2 has also been investigated. Figure 9 reports measurement results for RF injections performed on the four SPI lines at the same time and injection performed on the SCK (serial clock) SPI line only. The respective total RF induced voltage at the same pins is reported in Figure 10. In the first case the four SPI injection capacitors to the signal terminal of the SMA connector in Figure 2 are all connected. In the latter case, only one injection capacitor between the signal terminal of the SMA connector in Figure 2 and the SCK pin of the IC is mounted on the PCB.

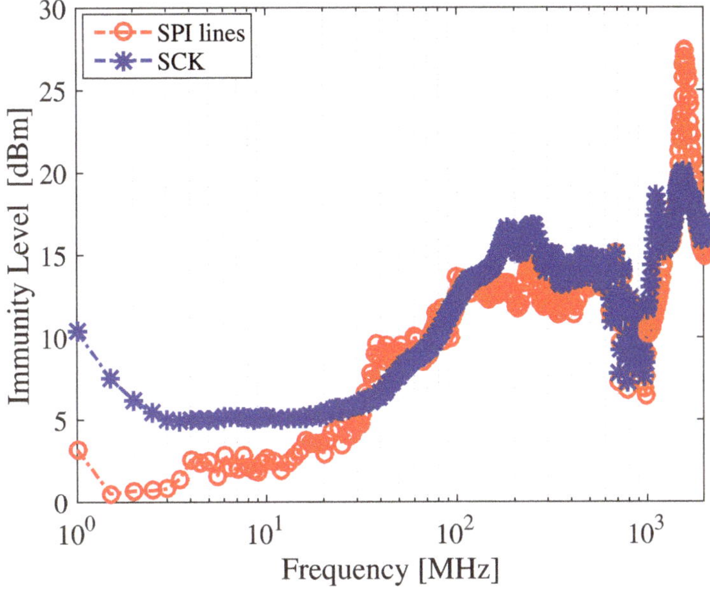

Figure 9. Measured immunity level for DPI on the SPI input pins: simultaneous injection on the four SPI lines and injection on the single SCLK line.

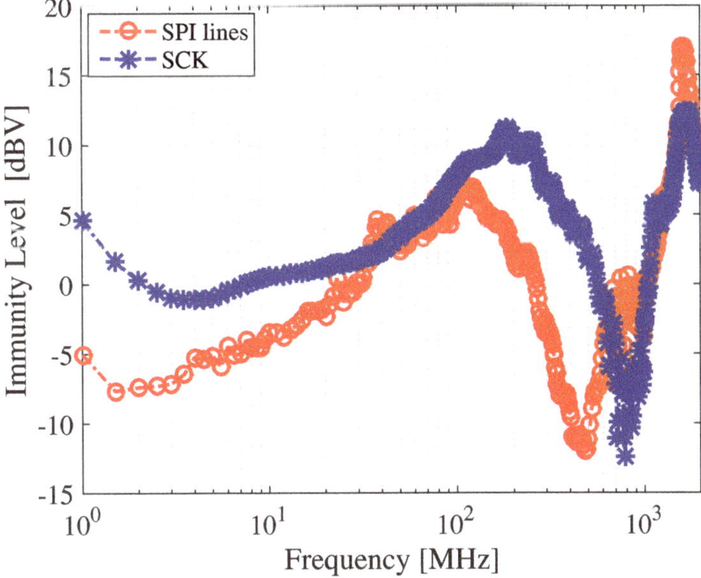

Figure 10. Total RF voltage induced at the SPI input pins: simultaneous injection on the four SPI lines and injection on the single SCLK line.

Failure Mechanisms

By comparing the immunity level reported in Figure 9 for the SPI injection with the immunity level reported in Figure 4 for cell inputs injection, it can be highlighted that the immunity level for SPI injection is lower than the immunity level of the unfiltered cell inputs. This is a serious concern, since the EMI immunity for SPI injection cannot be improved by filtering [16] because EMI filters would give rise to an unacceptable degradation of nominal digital waveforms.

An analysis of the mechanisms giving rise to EMI-induced failures when DPI is performed on SPI lines has been carried out on the basis of the data retrieved during the DPI tests, in analogy to what was discussed in Section 4.1.2. Based on these data, failures below 100 MHz are related, as one could expect, to errors in the SPI transmission detected by checking the PEC code. At higher frequency, however, EMI failures for SPI injection are related to an offset in the acquired cell voltages, equal for all the six cells, as shown in Figure 11. Such a behavior, which is similar to what was highlighted for DPI on the cell inputs in the high frequency range (Figure 6), can be related to EMI propagation inside the DUT to the internal ADC and/or to its reference voltage source.

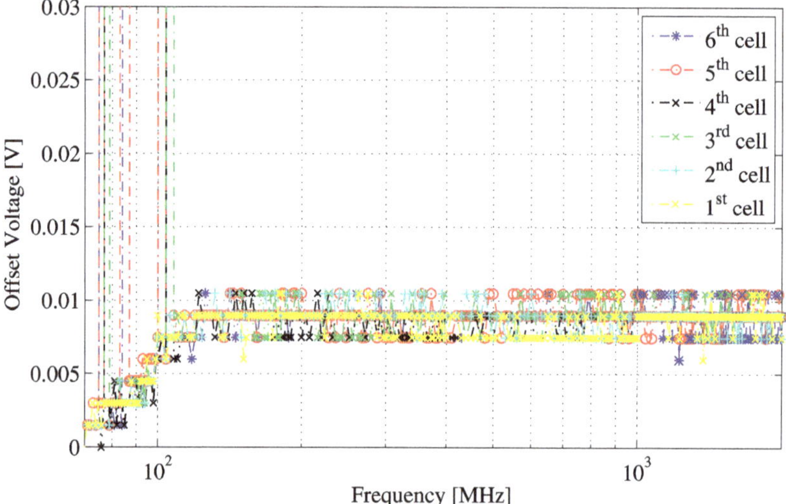

Figure 11. EMI-induced offset in the cell voltage readings obtained in accordance with failure levels of Figure 9 above 50 MHz (simultaneous injection on the four SPI pins).

5. Discussion: DPI Tests

DPI measurements have been performed in the bandwidth 1 MHz to 2 GHz up to a maximum incident power of 37 dBm on a commercial BMS. A test board has been specifically designed and fabricated in order to perform DPI in compliance with IEC 62132-4. In particular, DPI has been performed:

- On the cell monitoring inputs of the BMS IC connected to the top cell in the stack to be monitored (DM and CM injection), with and without RC low-pass filters recommended by the manufacturer;
- On the SPI lines.

During the tests, the following malfunctions in the operation of the BMS IC have been reported:

- Offset in the acquired cell voltages (considering that the target accuracy level of the IC an offset exceeding 10 mV has been considered as a failure);
- SPI communication failures (PEC failure and/or communication impaired).
- An increase in the current absorption from the power supply.

Offsets in acquired cell voltages have been reported both by DPI on the cell monitoring inputs and on the SPI communication lines. For what concerns cell monitoring input injection, an offset voltage has been reported for the cell undergoing injection, but also for other cells, depending on frequency. Communication failures have been reported for DPI in the communication lines only.

For what concerns DPI on cell monitoring inputs without the RC filters, a DPI immunity level from 5dBm to 15dBm has been reported up to 1 GHz. No substantial difference can be noticed for CM and DM injection. For what concerns DPI on cell monitoring inputs with the RC filters, a DPI immunity level exceeding the test level of 37 dBm has been reported for most frequencies above 5 MHz. Nonetheless, an immunity level as low as 20 dBm (for DM injection) has been reported in a frequency band around 700 MHz, in accordance with the parallel resonance of the filter capacitor (which is operating above its self-resonant frequency and shows an inductive impedance) and PCB parasitics. When filters were mounted, a slightly higher immunity level was measured for CM rather than for DM injection.

For what concerns DPI on SPI lines, a DPI immunity level lower than 5 dBm has been measured both at low frequency (<30 MHz) and in the bandwidth 700 MHz to 1 GHz. A slightly worse behavior has been reported for simultaneous injection on all the four SPI lines at the same time, rather than for injection performed on a single line. It is worth noting that unlike cells inputs, SPI communication lines cannot be filtered in order to avoid unacceptable degradation of the digital signal to be transmitted.

On the basis of the results of DPI tests, it can be observed that the immunity level of the BMS IC can be severely limited by its susceptibility to EMI superimposed onto the SPI line inputs.

6. Radiated Susceptibility Tests

In order to further investigate the BMS IC susceptibility, radiated EMI tests of a BMS IC have been performed in anechoic chamber in compliance with ISO11452-2 [26]. The DPI injection networks have been removed and the EMI filters on the cell inputs prescribed by the manufacturer have been included. The antenna has been placed both in front of the DUT and the cable harness. The DUT is located inside an anechoic chamber over a metal plane and it is remotely controlled and monitored by means of optical links. A sketch of the measurement setup is reported in Figure 12. The two terminals of the battery pack, monitored by the BMS, are connected through a 1.5 m-long cable to two LISNs. According to [26], the operation of the DUT has been tested radiating the DUT with a 200 V/m incident EM field, a typical test level for safety critical automotive applications, in the 200 MHz–1.4 GHz bandwidth, considering both horizontal and vertical polarization.

Figure 12 shows the radiated test setup with the antenna placed in front of the equipment under test (EUT corresponds to the PCB with the DUT and the battery pack). Referring to the this test setup, the EMI-induced wrong readings (area in gray) and the offset in the cell voltage readings in the anechoic chamber for vertical an horizontal polarizations are reported respectively in Figures 13 and 14. Both for vertical and horizontal polarization, the radiated field gives rise to completely wrong ADC readings, corresponding to the maximum or to the minimum values of the ADC range for a frequency lower than 400 MHz in the gray area in both Figures 13 and 14). The same phenomena are also observed in the range 850–950 MHz for horizontal polarization. Moreover, the results of radiated tests in Figures 13 and 14 highlight an EMI-induced offset in the acquired voltages in the range 650–900 MHz.

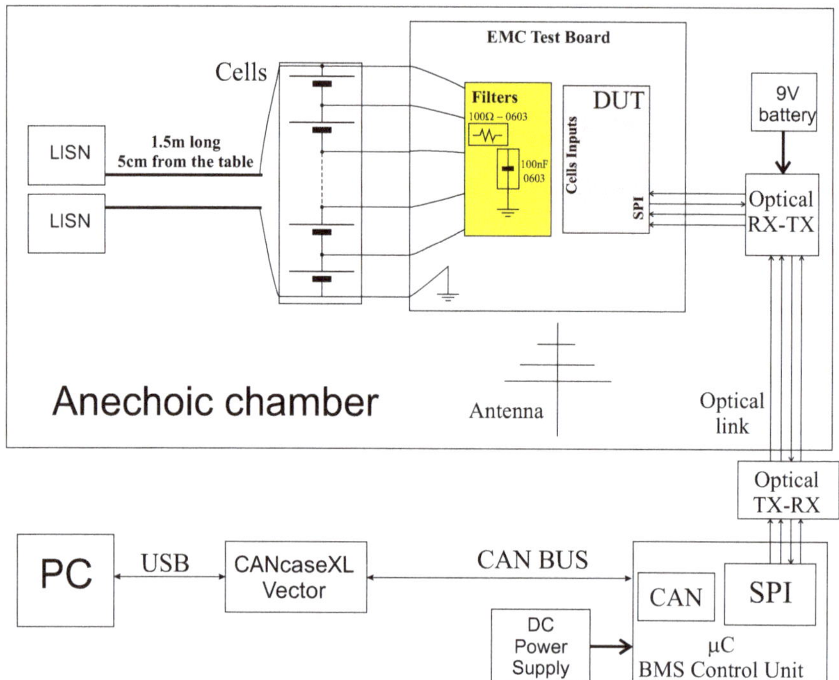

Figure 12. Radiated susceptibility test setup. Antenna in front of the equipment under test (EUT).

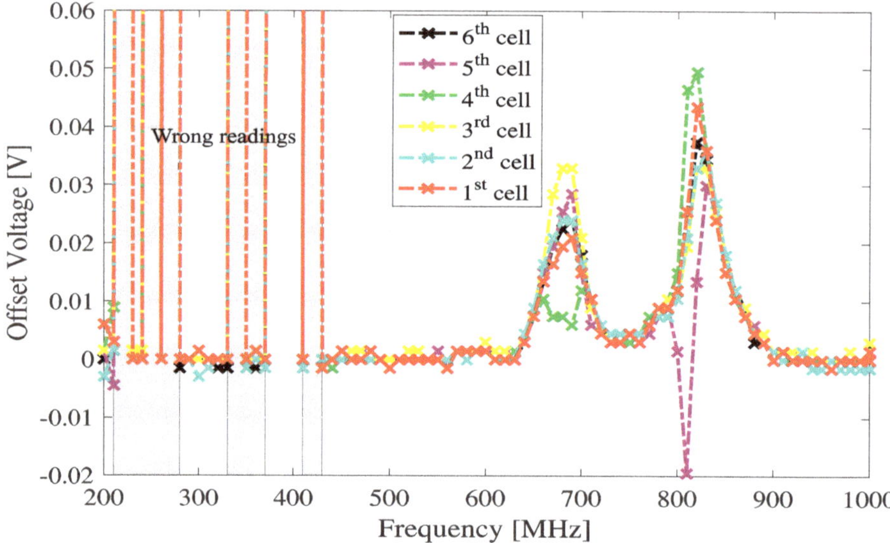

Figure 13. EMI-induced offset in the cell voltage readings. Antenna in front of the EUT (Figure 12). Vertical polarization.

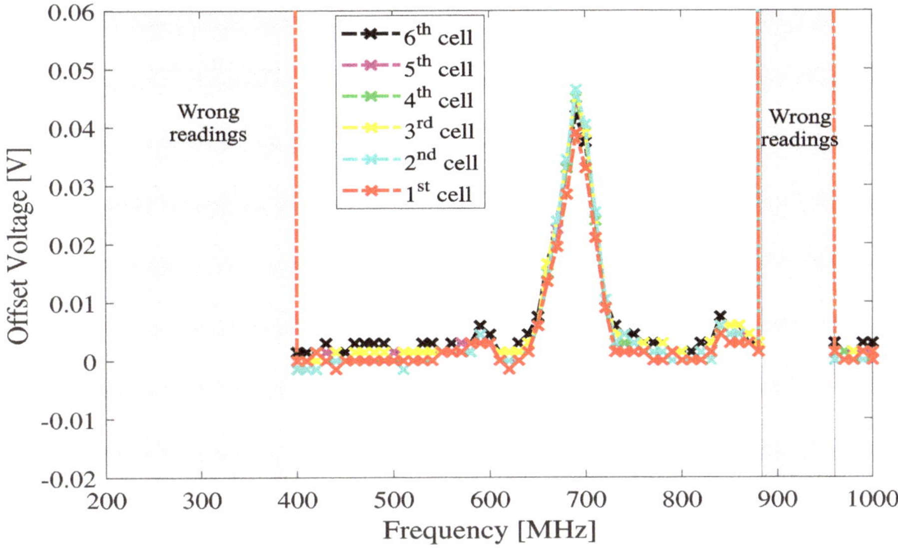

Figure 14. EMI-induced offset in the cell voltage readings. Antenna in front of the EUT (Figure 12). Horizontal polarization.

The same measurements have been repeated by moving the antenna in front of the cable, obtaining a similar phenomena. The respective wrong readings and EMI-induced offsets in the cell voltage readings in the anechoic chamber for vertical and horizontal polarizations are reported respectively in Figures 15 and 16.

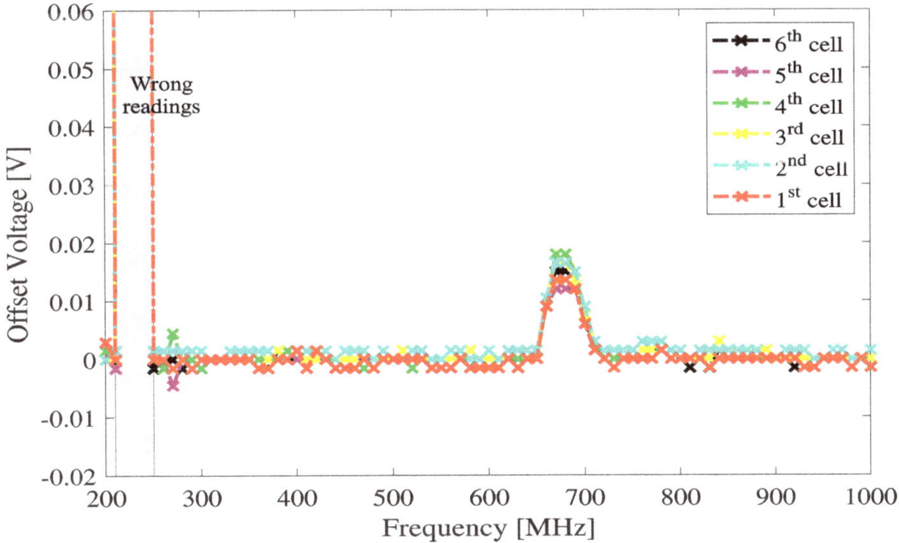

Figure 15. EMI-induced offset in the cell voltage readings. Antenna in front of the cable. Vertical polarization.

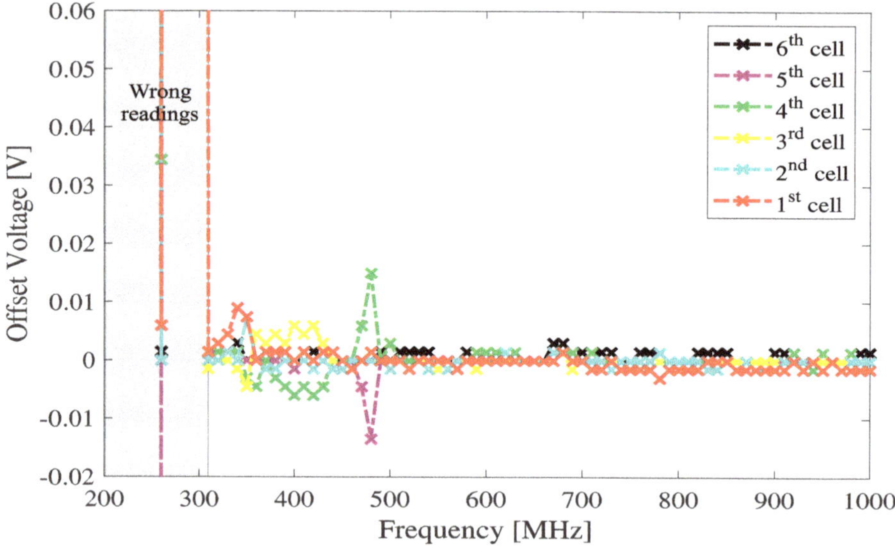

Figure 16. EMI-induced offset in the cell voltage readings. Antenna in front of the cable. Horizontal polarization.

7. Discussion: DPI vs. Radiated Immunity Tests

The radiated susceptibility measurements performed in compliance with ISO 11425-2 [26] aimed for establishing a correlation between the EMC performance of the BMS IC, previously assessed by the DPI method, and the susceptibility to EMI of a realistic BMS system IC. For a direct comparison, Figure 17 reports the immunity level measured by DPI tests on cell inputs (Figure 4)—its respective peak value in volts (from Figure 5)—and on the SPI lines (Figure 9); see Figure 17a–c respectively. Moreover, see the EMI-induced offsets in acquired cell voltages measured during radiated susceptibility tests in Figure 17d,e—for vertical and horizontal polarizations with the antenna placed in front of the EUT, respectively.

To obtain an approximate relation between DPI immunity level and radiated field strength, it has been observed that the failure threshold considered in DPI tests (EMI-induced offset in acquired cell voltages equal to 10 mV) has been reached irradiating the EUT in the bandwidth from 600 to 900 MHz by a vertically polarized E field $E_v = 100\,\text{V/m}$, which approximately induces a CM voltage on the BMS PCB lines connected to the cell inputs with a peak amplitude

$$V_{\text{cm}} = \int_0^h E_z(z)\text{d}z \simeq E_v \cdot h = 8.5\,\text{V}, \tag{4}$$

where $h = 8.5\,\text{cm}$ is the height of the harness connecting the BMS to the cells with respect to the ground plane. The EMI amplitude estimated by (4) is consistent with the results of the DPI immunity test reported in dBm and in volts as EMI peak amplitudes, respectively, in Figure 17a,b. In fact, in the bandwidth around 700 MHz where the effectiveness of RC filters is impaired by the resonance highlighted in (1), the EMI immunity level both for CM and DM injections is similar to that defined in (4).

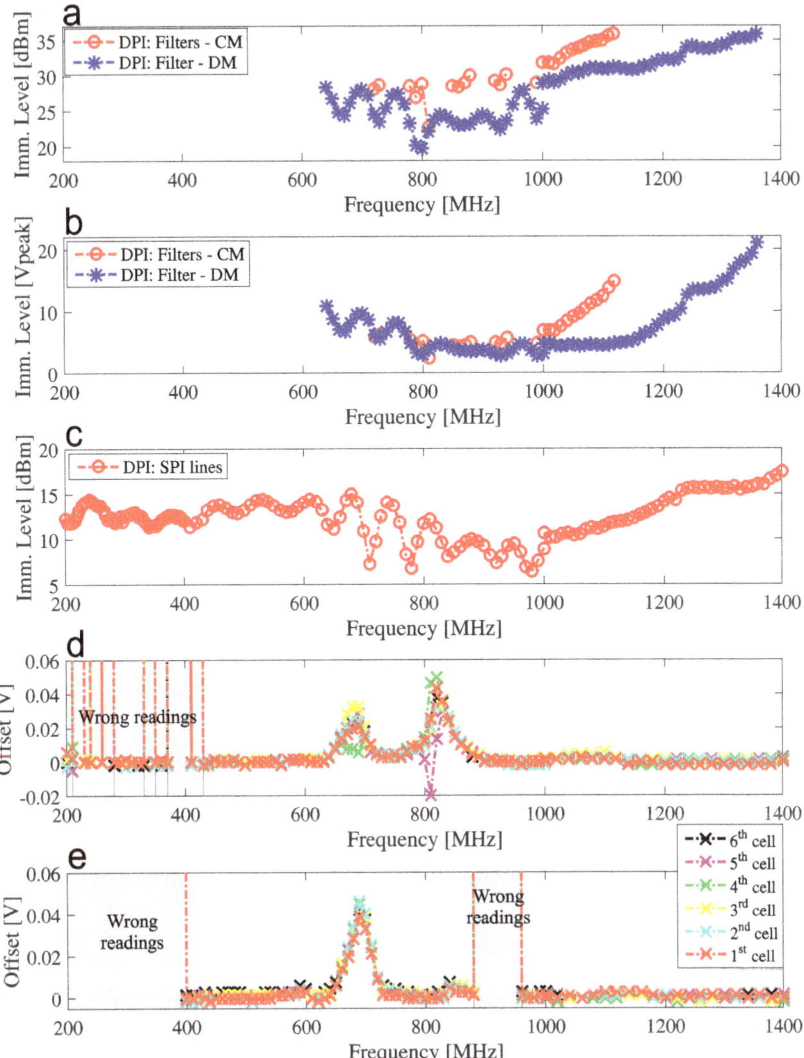

Figure 17. (**a**) Measured immunity level for DPI on the cell input pins: differential (DM) and common-mode (CM) injection, with 100 Ω, 100 nF RC filters in Figure 2, and (**b**) respective EMI peak amplitudes. (**c**) Measured immunity level for DPI on the four SPI lines. (**d**) EMI-induced offset in the acquired cell voltages, E = 200 V/m. Vertical polarization. (**e**) EMI-induced offset in the acquired cell voltages, E = 200 V/m, horizontal polarization.

Considering the significant differences of the injection mechanisms and of the test setup in DPI and radiated tests, further considerations on the correlation between IC-level DPI test results and system-level radiated immunity tests are difficult to be established and are also scarcely significant, since the results of radiated tests could be strongly influenced by the actual structure of the battery pack, including the BMS, which, in real applications, could be rather different with respect to the prototype considered in our investigation. Nonetheless, the same failure mechanisms (offset in acquired cell voltages and SPI communication failures) and the same critical EMI bandwidth highlighted during DPI tests have been found in system level tests.

In particular, an offset voltage of 10 mV, corresponding to the susceptibility level considered in DPI tests, has been observed by irradiating the system with an antenna placed in front of the EUT with a vertical polarization for a test frequency from 600–900 MHz. Considering that the board undergoing radiated tests includes the RC filters, such results are consistent with the results of DPI according to which the immunity of the EUT is particularly critical in the bandwidth around 700 MHz, where an immunity level of about 20 dBm was reported.

8. Conclusions

The susceptibility to EMI of a BMS front-end IC for EVs and HEVs has been investigated in this paper by DPI tests performed according with IEC 62132-4. On the basis of the experimental results, it has been highlighted that the BMS IC under test can be significantly susceptible to EMI injected on its cell input terminals and on its digital communication (SPI) lines. For what concerns cell inputs injection, in particular, it has been observed that low pass RC filtering can be effective at improving the immunity to EMI of the DUT in the 10–600 MHz bandwidth, but its effectiveness is reduced above 600 MHz. On the other hand, the susceptibility to EMI applied on the SPI input lines, which cannot be filtered, is likely to be a major concern for the specific application. Finally, the different mechanisms giving rise to EMI induced failures of the specific DUT have been highlighted. Depending on the EMI frequency and on the injection points, an EMI-induced offset in the cell readings, SPI failures and completely wrong acquired values have been reported. An abnormal current absorbtion from the cells while performing DPI has also been observed.

The immunity to EMI of the same BMS system has been addressed even by radiated susceptibility measurements performed in compliance with ISO 11425-2 to establish a correlation between the EMC performance of the BMS IC, previously assessed by the DPI method, and the susceptibility to EMI of a realistic BMS system based on the same IC. During the radiated tests, which have been performed for a field strength of 200 V/m, the same failure mechanisms highlighted during DPI tests have been observed (offset in acquired cell voltages and SPI communication failures).

Funding: The lab activities have been supported by the Regione Piemonte within the Biomethair Project, Fondo Europeo P.O.R. 2007–2013.

Acknowledgments: The author thanks P. Crovetti and F. Fiori from Politecnico di Torino, Italy, who supported a preliminary conference paper [4], and Eng. G. Borio, M. Scaglione, and A. Bertino of the LACE, Torino, Italy, for setting up the test bench used for the radiated emission measurements.

Conflicts of Interest: The author declares no conflict of interest.

References

1. Richelli, A.; Colalongo, L.; Kovács-Vajna, Z.M. Analog ICs for Automotive under EMI Attack. In Proceedings of the 2019 AEIT International Annual Conference, Florence, Italy, 18–20 September 2019; pp. 1–6.
2. Aiello, O.; Fiori, F. On the Susceptibility of Embedded Thermal Shutdown Circuit to Radio Frequency Interference. *IEEE Trans. EMC* **2012**, *54*, 405–412. [CrossRef]
3. Aiello, O. Hall-Effect Current Sensors Susceptibility to EMI: Experimental Study. *Electronics* **2019**, *8*, 1310. [CrossRef]
4. Aiello, O.; Crovetti, P.S.; Fiori, F. Susceptibility to EMI of a Battery Management System IC for electric vehicles. In Proceedings of the 2015 IEEE International Symposium on Electromagnetic Compatibility (EMC), Dresden, Germany, 16–22 August 2015; pp. 749–754.
5. Aiello, O.; Fiori, F. A New MagFET-Based Integrated Current Sensor Highly Immune to EMI. *Microelectron. Reliab.* **2013**, *53*, 573–581. [CrossRef]
6. Aiello, O.; Fiori, F. A new mirroring circuit for power MOS current sensing highly immune to EMI. *Sensors* **2013**, *13*, 1856–1871. [CrossRef] [PubMed]

7. Mutoh, N.; Nakanishi, M.; Kanesaki, M.; Nakashima, J. EMI noise control methods suitable for electric vehicle drive systems. *IEEE Trans. Electromagn. Compat.* **2005**, *47*, 930–937 [CrossRef]
8. Ansean, D.; García, V.M.; González, M.; Blanco-Viejo, C.; Viera, J.C.; Pulido, Y.F.; Sánchez, L. Lithium-Ion Battery Degradation Indicators Via Incremental Capacity Analysis. *IEEE Trans. Ind. Appl.* **2019**, *55*, 2991–3002. [CrossRef]
9. Omariba, Z.B.; Zhang, L.; Sun, D. Review on Health Management System for Lithium-Ion Batteries of Electric Vehicles. *Electronics* **2018**, *7*, 72. [CrossRef]
10. Lelie, M.; Braun, T.; Knips, M.; Nordmann, H.; Ringbeck, F.; Zappen, H.; Sauer, D.U. Battery Management System Hardware Concepts: An Overview. *Appl. Sci.* **2018**, *8*, 534. [CrossRef]
11. Chen, C.L.; Wang, D.S.; Li, J.J.; Wang, C.C. A Voltage Monitoring IC With HV Multiplexer and HV Transceiver for Battery Management Systems. *IEEE Trans. VLSI* **2015**, *23*, 244–253. [CrossRef]
12. Chol-Ho, K.; Moon-Young, K.; Gun-Woo, M. A Modularized Charge Equalizer Using a Battery Monitoring IC for Series-Connected Li-Ion Battery Strings in Electric Vehicles. *IEEE Trans. Power Electron.* **2013**, *28*, 3779–3787.
13. Cheng, K.W.E.; Divakar, B.P.; Wu, H.; Ding, K.; Ho, H.F. Battery-Management System (BMS) and SOC Development for Electrical Vehicles. *IEEE Trans. Veh. Technol.* **2011**, *60*, 76–88. [CrossRef]
14. Spadacini, G.; Pignari, S.A. Numerical Assessment of Radiated Susceptibility of Twisted-Wire Pairs With Random Nonuniform Twisting. *IEEE Trans. EMC* **2013**, *55*, 956–964. [CrossRef]
15. Richelli, A. EMI Susceptibility Issue in Analog Front-End for Sensor Applications. *J. Sens.* **2016**, *2016*, 1082454. [CrossRef]
16. Crovetti, P.; Fiori, F. IC digital input highly immune to EMI. In Proceedings of the 2013 International Conference on Electromagnetics in Advanced Applications (ICEAA), Torino, Italy, 9–13 September 2013; pp. 1500–1503.
17. Richelli, A.; Matiga, G.; Redoute, J.M. Design of a Folded Cascode Opamp with Increased Immunity to Conducted Electromagnetic Interference" in 0.18 um. *Microelectron. Reliab.* **2015**, *55*, 654–661. [CrossRef]
18. Richelli, A.; Delaini, G.; Grassi, M.; Redoute, J.M. Susceptibility of Operational Amplifiers to Conducted EMI Injected Through the Ground Plane into Their Output Terminal. *IEEE Trans. Reliab.* **2016**, *65*, 1369–1379. [CrossRef]
19. Andrea, D. *Battery Management Systems for Large Lithium-Ion Battery Packs*; Artech House Inc: Boston, MA, USA, 2010.
20. Jiang, J.; Zhang, C. *Fundamentals and Applications of Lithium-Ion Batteries in Electric Drive Vehicles*; John Wiley & Sons Singapore Pte. Ltd.: Singapore, 2015.
21. Hauser, A.; Kuhn, R. 12—Cell balancing, battery state estimation, and safety aspects of battery management systems for electric vehicles. In *Advances in Battery Technologies for Electric Vehicles*; Scrosati, B., Garche, J., Tillmetz, W., Eds.; Woodhead Publishing: Cambridge, UK, 2015; pp. 283–326.
22. Lu, L.; Han, X.; Li, J.; Hua, J.; Ouyang, M. A review on the key issues for lithiumion battery management in electric vehicles. *J. Power Sources* **2013**, *226*, 272–288. [CrossRef]
23. Doughty, D.; Roth, P. A general discussion of Li-ion battery safety. *Electrochem. Soc. Interface* **2012**, *21*, 37–44.
24. IEC 623132-4:2006. Integrated Circuits, Measurement of Electromagnetic Immunity—Part 4: Direct RF Power Injection Method. Available online: https://webstore.iec.ch/publication/6510 (accessed on 2 February 2020).
25. Generic IC EMC Test Specification, Ed. 2.1. 2017. Available online: http://www.zvei.org/fileadmin/user_upload/Presse_und_Medien/Publikationen/2017/Juli/Generic_IC_EMC_Test_Specification/Generic_IC_EMC_Test_Specification_2.1_180713_ZVEI.pdf (accessed on 19 January 2020).
26. Road Vehicles—Vehicle Test Methods for Electrical Disturbances from Narrowband Radiated Electromagnetic Energy, ISO Std. 11451-2. 2005. Available online: https://www.iso.org/standard/37999.html (accessed on 2 February 2020).

Article

EMI Susceptibility of the Output Pin in CMOS Amplifiers

Anna Richelli *, Luigi Colalongo and Zsolt Kovacs-Vajna

Department of Information Engineering, University of Brescia, 25123 Brescia BS, Italy; luigi.colalongo@unibs.it (L.C.); zsolt.kovacsvajna@unibs.it (Z.K.-V.)

* Correspondence: anna.richelli@unibs.it; Tel.: +39-030-371-5501

Received: 16 January 2020; Accepted: 5 February 2020; Published: 9 February 2020

Abstract: Measurements in commercial devices demonstrate a considerable susceptibility of the operational amplifiers to the electromagnetic interferences coupled to their output pin. This paper investigates some basic architectures starting from single stage amplifiers up to a whole operational amplifier. The result is a correlation between the different amplifier configurations, the output impedance and the susceptibility to the interferences. The simulations are perfomed by using the standard CMOS UMC 180nm technology and by running the netlist of the schematics extracted from the layout.

Keywords: Electromagnetic Interferences; Amplifiers; CMOS integrated circuits; susceptibility of the output pin

1. Introduction

The assessment of IC immunity to EMI is presently compulsory for many applications, because, thanks to the progress of VLSI, electronics is ubiquitous. It is evident in the following industries: computing, communication, manufacturing, transport, entertainment, business, health. We can find ICs nearly everywhere: in computers (of course), in cellphones, watches, medical devices, cars.... Moreover, in recent years, raises serious preoccupations on the vulnerability of electronic equipment to intentional EMI attacks [1–5]: sources capable of producing very high RF power over a wide frequency range are, indeed, presently easily available or even home-made manufacturable. In general, as shown in Figures 1 and 2, Electromagnetic Compatibility issues can be divided into two parts: emissions and susceptibility. Both of them should be as low as possible. This paper will focus on susceptibility, in particular of the analog ICs to conducted interferences. Due to the small dimensions of Integrated Circuits indeed, susceptibility and emission via conduction are more important than via radiation. The susceptibility of analog integrated circuits to conducted EMI has been modelled and verified with simulations as well as experimentally in several works, and analytical models predicting high-power EMI effects in CMOS operational amplifiers and offering a good matching with experimental results have been presented in [6]. On the other hand, the exact repercussion of electromagnetic interference (EMI), which is injected conductively into arbitrary pins of an integrated circuit (IC), is very hard to predict owing to the fact that all the existing coupling paths need to be taken into account [7]. Several works can be found in the literature, describing the susceptibility of analog ICs, as these are more sensitive to EMI. Most of this research focuses on operational amplifiers and other widely used analog blocks, such as for example voltage references, because of their high sensitivity to EMI and their prevalence in electronic circuits [8–15]. The EMI signals are considered directly injected into the gate of the differential pair or into the power supply pins (Vdd and Vss). On the other hand, all the pins can be a possible injection point as demonstrated, for example in [16], where the susceptibility of a bulk-driven CMOS amplifier is investigated by injecting the interferences into the bulk pin. Moreover, the direct injection of the interferences is not the only path available for the EMI. Indeed in [17], the interferences reach the IC (a Local Interconnect Network (LIN) integrated output driver) by means of a parasitic capacitive coupling between the output pin and the conductive ground plane. The unwanted RF signals may indeed be transmitted through the copper plane, often shared with other analog, digital or mixed ICs, and can be capacitively coupled to all the pins, including the output one [18–20].

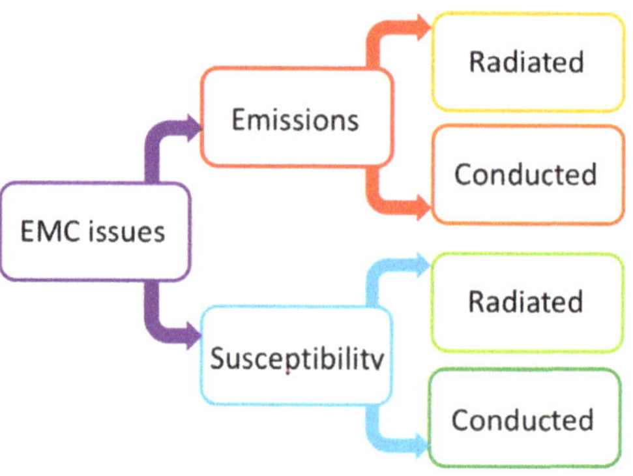

Figure 1. EMC issues.

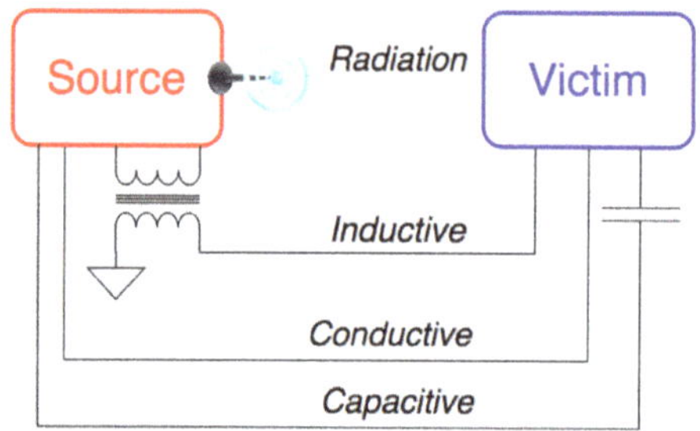

Figure 2. Possible coupling mechanisms for EMI attack.

Precisely, the susceptibility of the Operational Amplifiers to the Electromagnetic Interferences capacitively coupled to the ouput pin has been demonstrated in recent works [18–20], involving both commercial amplifiers and custom CMOS integrated ones. The results show a considerable vulnerability and exacerbate the scenario of EMI pollution. Moreover, it follows that the PCB ground plane, commonly shared with other analog, digital or mixed-signal circuits, can be a critical point of EMI pickup and injection. For the sake of clarity, the most relevant measurement results, provided by the literature [18,19], are plotted in Figures 3–8. The first four graphics show the EMI induced offset in well-known commercial amplifiers (ua741, OPA705, NE5534 and ICL7611). We can see that regardless the techonology and the architecture of the amplifiers (bipolar, CMOS or BiCMOS), they are all suscpetible to the interferences coupled to the output pin. Also, the offset is very high around the hundreds of MHz and it depends on the amplitude of the interferences. The second couple of figures refers to some commercial voltage references. They are very sensible to the EMI coupled to the output pin.

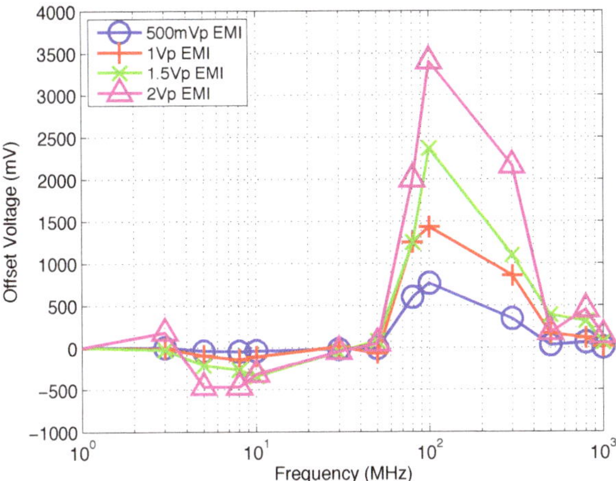

Figure 3. EMI induced offset in ua741: interferences coupled to the output pin.

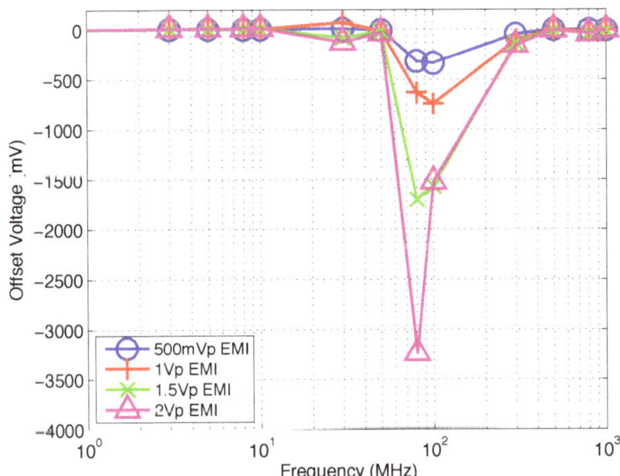

Figure 4. EMI induced offset in OPA705: interferences coupled to the output pin.

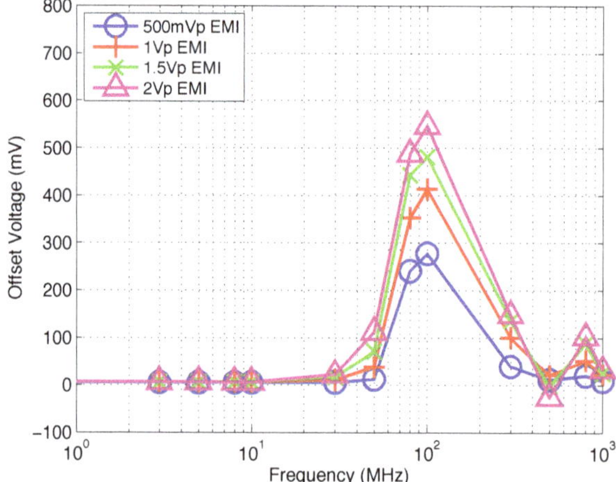

Figure 5. EMI induced offset in NE5534: interferences coupled to the output pin.

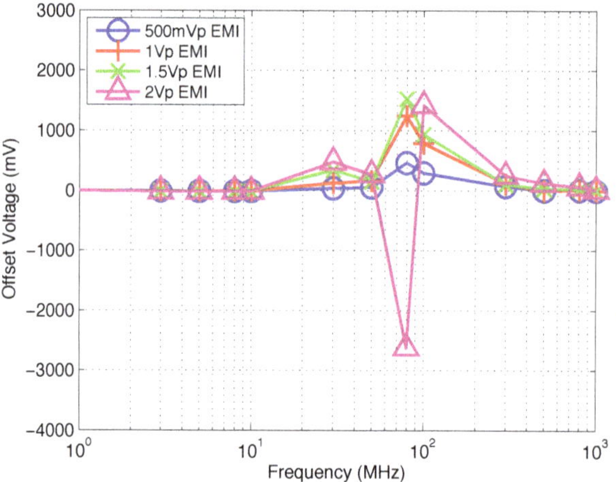

Figure 6. EMI induced offset in ICL7611: interferences coupled to the output pin.

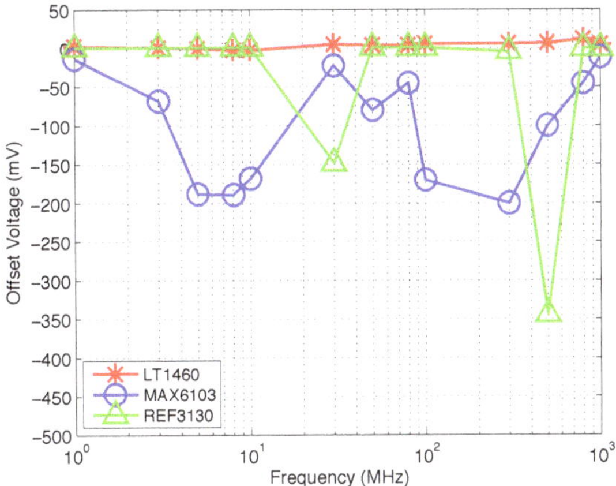

Figure 7. EMI induced offset in series voltage references: interferences coupled to the output pin.

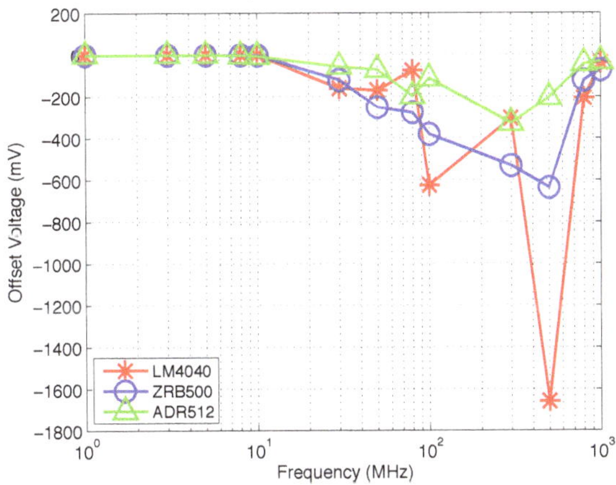

Figure 8. EMI induced offset in shunt voltage references: interferences coupled to the output pin.

Therefore, it is of paramount importance to deeply understand the reason of the susceptibility, the possible injection paths and the differences between the amplifier configurations. This paper has been organized as follows. In the second section, an overview about the simulation settings, the amplifier configurations and the transistors sizing will be provided; in the following section, a single stage common source amplifier will be studied when subjected to the EMI coupled to its output node (i.e., the drain terminal); in the fourth, fifth and sixth sections the same will be done for the other simple topologies (common gate, common drain and cascode, respectively); in Section 7 a two stages differential input Miller amplifier will be analyzed both in open loop and closed loop configurations. Finally, in Section 8, the discussion of the main results will be drawn and Section 9 concludes the paper.

2. Definition of the Simulation Settings

To reproduce the EMI pollution coming from the PCB, or from a generic ground plane or from close high-frequencies signal traces, an ideal sinusoidal voltage source is connected to the output pin of the generic amplifier, by means of an ideal capacitor, representing the capacitive coupling. A few picofarad forms a realistic value of the parasitic capacitance between a middle length trace and the PCB ground plane, as stated in [21–23]. The amplifiers are designed in a standard CMOS technology (the UMC 180nm) and the simulations are performed using the software Virtuoso, developed by Cadence, and, in particular, the Analog Design Environment (ADE) tool. In this way, the investigation of the EMI effect can be carried out with the biggest flexibility. Indeed, the sizing of the transistors can be easy changed, all the possible configuration of the amplifier can be simulated and the effect of the amplitude and the frequency of the interferences can be easy checked. The experimental measurements offer more eminent and complete results, but in this technology the simulations and the measurements of susceptibility fit rather well, Ref. [18,20,24]. Moreover, the simulations (if well supported by the transistor models) are a powerful instrument to understand the possible reasons of the suscepetibility. They indeed allow changing the conditions, the sizing, the settings, and so on..., and allow analyzing the effect of every change individually. Here, a list of the amplifier stages subjected to the interferences is briefly depicted. The DC and AC parameters and the transistors sizing are summarized in the following tables. The details on the circuit architectures, the schematics and the behavior of the amplifiers when subjected to the EMI are, instead, discussed in the corresponding sections. Table 1 refers to the common source stage shown in Figure 9; Table 2 summarizes the parameters of the common drain stage shown in Figure 10; in Table 3 the parameters of the cascode stage shown in Figure 11 are listed.

Table 1. DC and AC parameters and Transistor Sizing of the Common Source Stage.

	DC Biasing and AC Parameters	Components Sizing
Common Source with R (passive load)	VbiasN = 500 mV, 16 dB gain 1.5 MHz cut-off freq., 11 MHz GBW	W = 20 µm, L = 1 µm R = 10 kΩ
Common Source with PMOS (active load)	VbiasN = 500 mV, VbiasP = 1 V, 35 dB gain 180 kHz cut-off freq., 10 MHz GBW	Wn = 20 µm, Ln = 1 µm Wp = 20 µm, Lp = 1 µm

Table 2. DC and AC parameters and Transistor Sizing of the Common Drain Stage.

	DC Biasing and AC Parameters	Components Sizing
Common Drain with R (passive load)	VbiasN = 1.2 V, −1dB gain 13 MHz cut-off freq.	W = 20 µm, L = 1 µm R = 10 kΩ
Common Drain with NMOS (active load)	VbiasN = 1.2 and VbiasN2 = 500 mV, 0 dB gain 9.6 MHz cut-off freq.	Wn = 20 µm, Ln = 1 µm Wn2 = 20 µm, Ln2 = 1 µm

Table 3. DC and AC parameters and Transistor Sizing of the Cascode Stage.

	DC Biasing and AC Parameters	Components Sizing
Cascode with R (passive load)	VbiasN1 = 0.5 V, VbiasN2 = 1 V, 16 dB gain 1.5 MHz cut-off freq., 10 MHz GBW	Wn1,2 = 20 µm, Ln1,2 = 1 µm R = 10 kΩ
Common Drain with NMOS (active load)	VbiasN1 = 0.5 V, VbiasN2 = 1 V, VbiasP = 1.015 V 45 dB gain, 0.6 MHz cut-off freq., 10 MHz GBW	Wn1,2 = 20 µm, Ln1,2 = 1 µm Wp = 20 µm, Lp = 1 µm

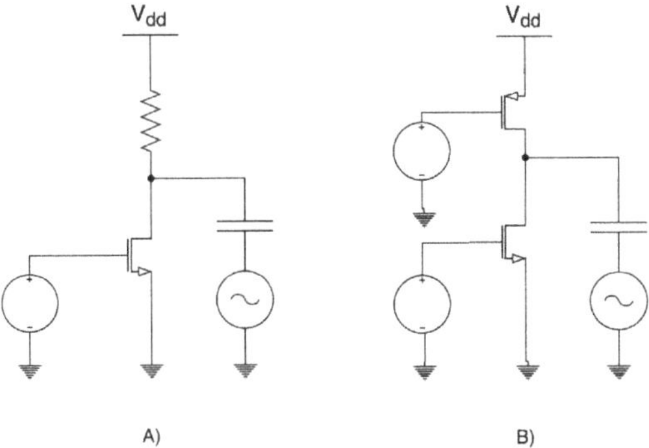

Figure 9. Schematics of the common source stage, with passive (**A**) and active (**B**) load respectively.

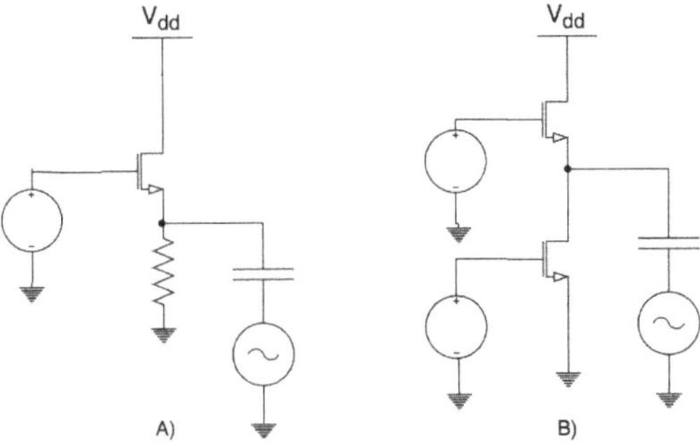

Figure 10. Schematics of the common drain stage, with passive (**A**) and active (**B**) load respectively.

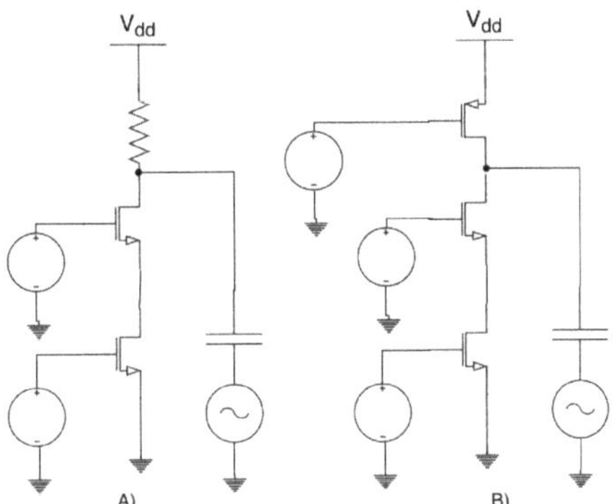

Figure 11. Schematics of the cascode stage, with passive (**A**) and active (**B**) load respectively.

3. Susceptibility of the Common Source Stage

As a first step, a common source stage (CS) is investigated. The CS amplifier is based on a NMOS transistor, designed in the standard CMOS UMC (United Microelectronics Corporation) 180 nm technology; it is a N_18_MM device (which means regular threshold transistor for 1.8 V nominal voltage supply), with aspect ratio 20, length 1 μm and width 20 μm. The biasing point of the device is given by an ideal voltage source of 500 mV at the transistor gate and a resistor of 10 kΩ connected between the drain and the Vdd (1.8 V), as shown in Figure 9A. The resistor acts also as the active load of the CS stage. With the threshold of 374 mV, the biasing point sets the current at 57 μA and these small signal parameters: gm 690 μS and gds 7.2 μS. As a single stage amplifier it exhibits a gain of 16 dB, a cut-off frequency of 1.5 MHz and a gain bandwidth product (GBW) of 11 MHz. The interference is represented by means of a sinusoidal signal with large amplitude and medium-large frequency. In particular, the amplitude is 1 V peak to peak and the frequency is ranging between 10 kHz and 10 GHz. The capacitive coupling versus the noisy ground plane is set to 1 pF. In this case, the comomon source amplifier exhibits a very low EMI induced offset (a few mV at maximum) and therefore a high immunity to the interferences coupled to the output pin. If a PMOS transistor takes the place of the resistor, as active load, the immunity changes. The new schematic is shown in Figure 9B. The resistive load has been substituted by an active load made of a PMOS transistor in saturation region, with the same aspect ratio of the NMOS counterpart. The gate of the PMOS is biased at 1V and the threshold voltage is around 500 mV. The operating point is slightly different to that of the first circuit and the current is now 54 μA. The small signal simulations show a gain of 35 dB, a cut-off frequency of 180 kHz and a GBW of 10 MHz.

In this second case (with active load), the common source stage presents a large susceptibility to the interferences coupled to its output pin. Indeed, the EMI induced offset on the output voltage reaches about 200 mV and it is rather large in a wide frequency range, starting from 500 kHz (offset is 100 mV) up to 10 GHz. The results of the simulations performed on both the amplifiers are shown in Figure 12. If the amplitude of interfering signal increases, the induced offset has a large raise, as expected. For example, if the EMI amplitude is 1 V, the maximum offset becomes 400 mV.

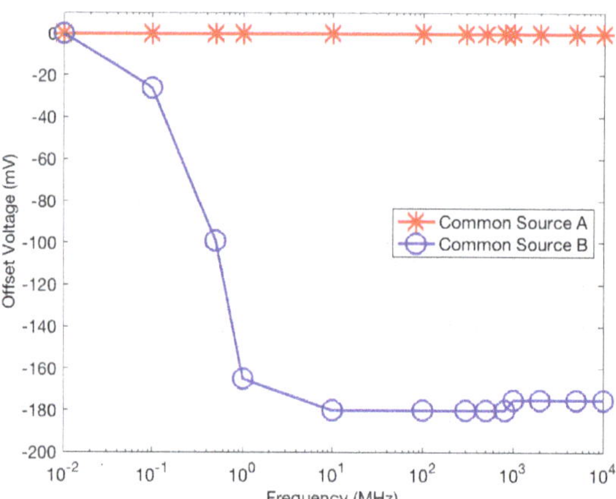

Figure 12. EMI induced offset in the common source stage.

4. Susceptibility of the Common Gate Stage

The behavior of the common gate stage (shown in Figure 13) is exactly the same of the common source stage. This is because the injection point (the drain of the transistor) is the same.

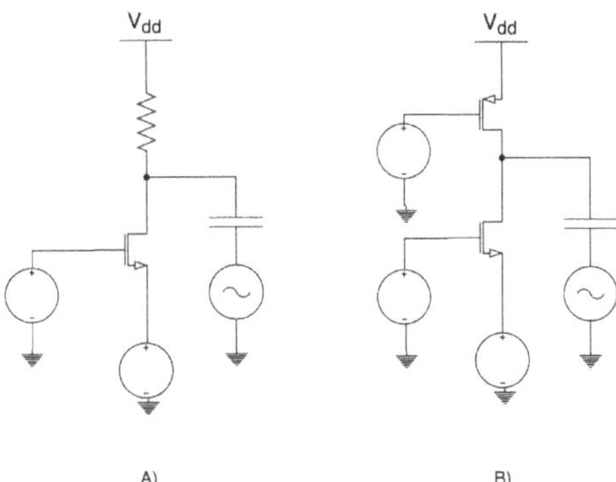

Figure 13. Schematics of the common gate stage, with passive (**A**) and active (**B**) load respectively.

5. Susceptibility of the Common Drain Stage

The behavior of the common drain stage is now investigated. The first case is the schematic shown in Figure 10A, with the resistive load.

The NMOS device is a regular threshold transistor, with aspect ratio 20, length 1μm and width 20 μm. The biasing point of the device is given by an ideal voltage source of 1.2 V at the transistor gate and a resistor of 10 kΩ connected between the source and gnd acts as a passive load. In this operating point the small signal parameters are: gm 763 μS and gds 8.2 μS. The current is 68 μA and the threshold voltage is 379 mV. The AC gain is −1 dB and the cut-off frequency is 13 MHz. The second case is the common drain with the active load (as in Figure 10B). The input transistor is biased at the same 1.2 V, while the active load NMOS is biased by an ideal voltage source of 0.5 V connected to its gate. The operating point is slightly different from that of case A: the current is now indeed 50 μA. The small signal parameters of the input transistor are now: gm 635 μS and gds 6.8 μS. The active load NMOS has similar parameters: gm of 633 μS and gds of 7.1 μS. The AC gain is 0 dB and the cut-off frequency is 9.6 MHz. To evaluate the susceptibility of the common drain stage the interferences are coupled to the transistor source (which is the output pin in this configuration) by means of a 1 pF capacitor. The EMI amplitude is 1 V peak to peak and the frequency is ranging between 10 kHz and 10 GHz. The results of the simulations performed on both the amplifiers are shown in Figure 14. As one can see both the schematics (A and B, with resistive or active load) are rather susceptible to the interferences, starting from a few MHz of the interfering signal frequency.

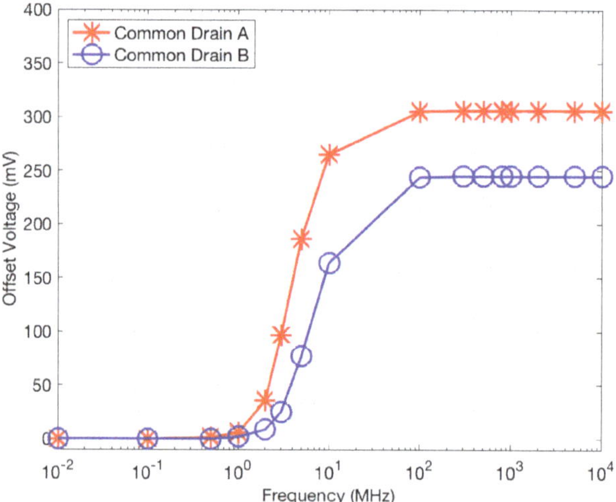

Figure 14. EMI induced offset in the common drain stage.

6. Susceptibility of the Cascode Stage

To have a complete overview of the single stage susceptibility, a cascode stage (common source + common gate) is also considered. The schematic is shown in Figure 11 with passive resistive load (A) and with active load (B). The transistors have always a length of 1 µm and a width of 20 µm. The common source transistor is biased with 0.5 V at the gate, while the common gate has a biased of 1V. The PMOS transistor in the schematic B is biased at 1.015 V in order to have operating points very similar between the circuits A and B, with the same current of about 50 µA and the same small signal parameters. In particular, the common source transistor has: gm 633 µS and gds 7.1 µS. The common gate transistor has: gm 633 µS and gds 7.4 µS. Finally, the active load PMOS transistor has: gm 301 µS and gds 3.5 µS. In the first case (schematic A), the AC gain is 16 dB, the cut-off frequency is 1.5 MHz and the GBW is around 10 MHz. In the second case (schematic B), the AC gain is 45 dB, the cut-off frequency is 0.6 MHz and the GBW is around 10 MHz. The behavior of the cascode stage seems very similar to that of the common source stage. The circuit with passive load presents a negligible EMI-induced offset, while the circuit with active load is rather susceptible. The results of the simulations performed on both the amplifiers are shown in Figure 15.

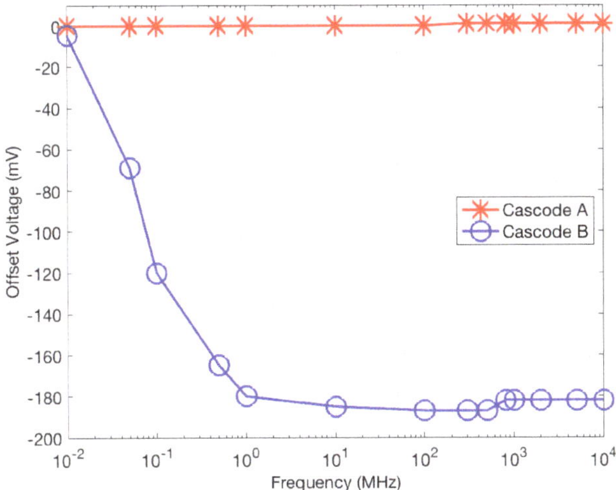

Figure 15. EMI induced offset in the cascode stage.

7. Susceptibility of a Two Stages Amplifier

The next step is to investigate the susceptibility of a two stages differential input single ended amplifier. Therefore, a Miller amplifier has been considered. The Miller amplifier is indeed one of the most common CMOS OpAmp. It is based on a differential input stage with single ended conversion which has a behavior similar to that of a common source stage. The second gain stage is usually a common source plus an active load. Sometimes, depending on the applications, a voltage buffer is added after the second stage, to have a low output impedance.

7.1. Susceptibility of a Two Stages Amplifier in Open Loop Configuration

In this subsection, we consider a two stages amplifier in an open loop. This is usually the case of an amplifier used as a comparator. The schematics of the circuits are shown in Figure 16. On the left (Figure A) there is the amplifier with a resistive load, while on the right (Figure B) there is the same amplifier with active load. Concerning the A circuit, the AC simulations show a gain of 55 dB, a cut-off frequency of 22 kHz, a GBW of 12 MHz. Regarding the B circuit, the gain is 72 dB (thanks to the active load), the cut-off frequency is 2.8 kHz and the GBW is 12 MHz. Although both the circuits are now used in an open loop configuration, they have a frequency compensation RC network that ensures 75° of phase margins if the circuits are in a closed loop. The main small signal parameters and the operating point are also listed in Table 4.

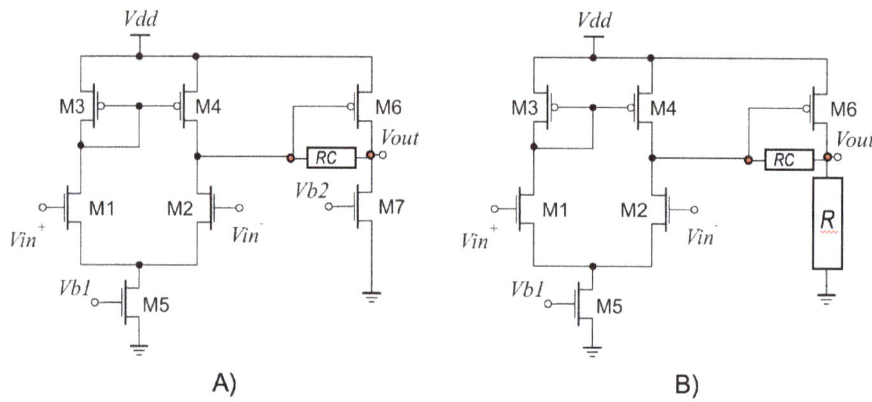

Figure 16. Schematic ofthe Miller amplifier: (**A**) with active load, (**B**) with resistive load.

Table 4. DC and AC parameters and Transistor Sizing of the Two Stages Amplifier.

	DC Biasing and AC Parameters	Components Sizing
Miller with R (passive load)	Vb1 = 1.16 V, 55 dB gain 22 kHz cut-off freq., 12 MHz GBW	Wn1,2 = 20 μm, Ln1,2 = 1 μm; Wp3,4,6 = 25 μm, Lp3,4,6 = 0.5 μm Wn5 = 20 μm, Ln5 = 1 μm; R = 10 kΩ
Miller with NMOS (active load)	Vb1 = Vb2 = 1.16 V, 72 dB gain, 2.8 kHz cut-off freq., 12 MHz GBW	Wn1,2 = 20 μm, Ln1,2 = 1 μm; Wp3,4,6 = 25 μm, Lp3,4,6 = 0.5 μm Wn5 = 20 μm, Ln5 = 1 μm; Wn7 = 8.5 μm, Ln7 = 0.5 μm

The offset induced by the interfering signals capacitively coupled to the output pin is shown in Figure 17: the blue line represents the Miller amplifier with active load (A circuit in Figure 16) and the red one represents the amplifier with the passive load (B circuit in Figure 16). The output stage of the amplifier in the open loop configuration is exactly a common source stage and a similar EMI susceptibility could be expected. Instead, as shown by the simulation results plotted in Figure 17, the EMI induced offset has two main differences. The first one is the susceptibility of the amplifier with passive load: although it is much less than the amplifier with active load, it is rather high, much higher than the susceptibility of the simple common source stage with passive load (which is negligible). The maximum offset is indeed 600 mV and the critical frequencies range from tens of MHz up to Ghz. The second difference is that the susceptibility of the amplifier with active low decreases at very high frequencies. It does not happen in the simple stages (common source, gate, drain and cascode).

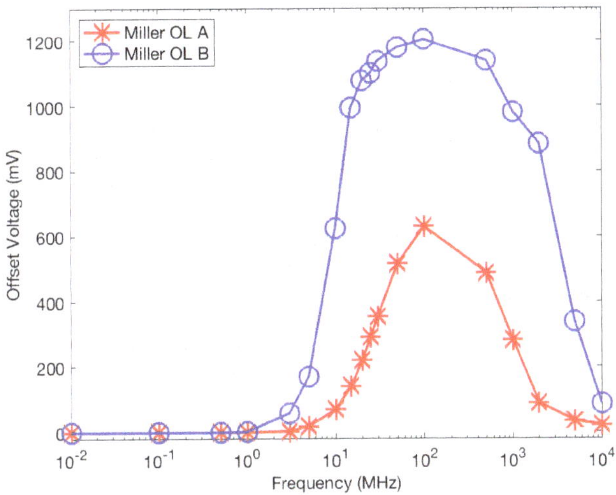

Figure 17. EMI induced offset in the Miller amplifier, open loop configuration.

Finally, another huge difference is the value of the offset, which is much larger than that of the single stage: indeed, the maximum offset is above 1 V in the Miller amplifier, while it is less than 200 mV in the common source stage. One of the possible reason of the different behavior could be the RC network for the frequency compensation. It is indeed in the amplifier schematic, even if it is used in an open loop. The sizing of the RC network is: 3 kΩ for the resistor and 1 pF for the capacitor. The RC network is directly connected to the output pin, it is across the first and the second stage and it may cause a typical phenomenon, Ref. [25–27], called EMI charge pumping, which leads to a severe DC shift of the biasing point. To investigate this, a schematic without the RC network (see Figure 18) has been simulated and the simulation results are plotted in Figure 19.

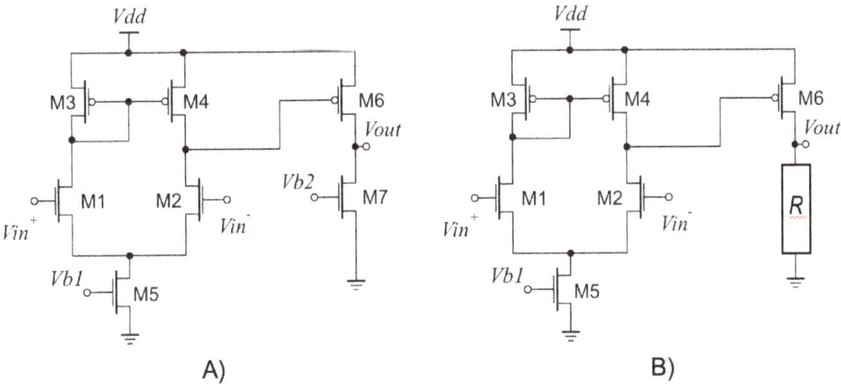

Figure 18. Schematic ofthe Miller amplifier without the RC network for the frequency compensation: (**A**) with active load, (**B**) with resistive load.

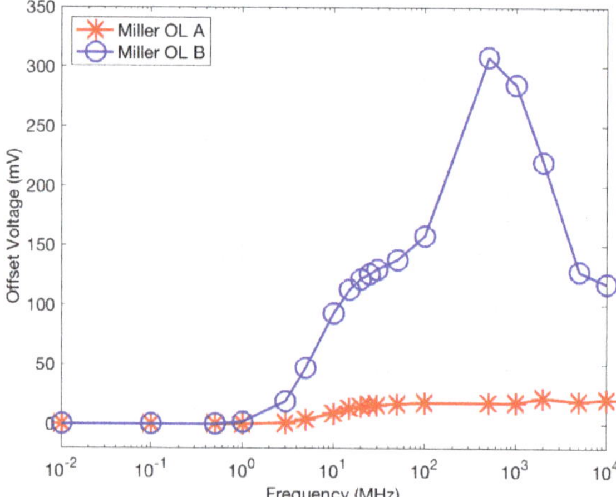

Figure 19. EMI induced offset in the Miller amplifier without RC network, open loop configuration.

By comparing Figures 17 and 19, it results that: the EMI induced offset is much reduced, and it is almost negligible in the case of resistive load (B circuit). On the other hand, there are still some differences which can be due to the first stage of the amplifier, which polarizes the second stage and is also weakly connected to the ouput node by means of the parasitics.

7.2. Susceptibility of a Two Stages Amplifier in Closed Loop Configuration

The circuits shown in Figure 16 are now connected in a closed loop and the effect of interferences are investigated. The feedback loop decreases the output impedance and a reduction of the EMI induced offset is reasonably expected. The results are indeed plotted in Figure 20.

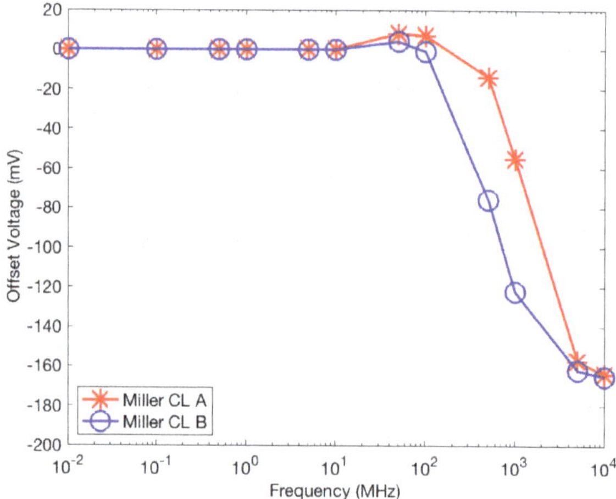

Figure 20. EMI induced offset in the Miller amplifier with RC network, closed loop configuration.

The susceptibility is similar to that of a simple common source stage (the maximum EMI induced offset is the same); the behavior of the amplifier with the active load is very similar to that of the amplifier with resistive load and the range of uncritical frequencies is enlarged up to a few hundreds of MHz.

8. Overview of the Obtained Results

The exact repercussion of electromagnetic interference, which is injected into arbitrary pins of an integrated amplifier, is very hard to predict owing to the fact that all the existing coupling paths need to be taken into account.

The investigation of single stage amplifier, such as common source, common gate, common drain and cascode, is easier, but non trivial. The susceptibility can be related, indeed, to the output impedance. The common source, common gate and cascode stage have a high impedance output node and it is known that the high impedance (floating) node are more prone to pick up the noise. The behavior of the common drain stage is less intuitive, but it can be still attributed to the impedance: at large frequencies the output impedance of the common drain (which is small at low frequencies) grows up. Moreover, the parasitic gate-source capacitance of the common drain creates a feedback from the output to the input and the charge pumping is maximum. This explains the increasing of the EMI induced offset above the hundreds of MHz.

Regarding the two stages amplifier, a distinction must be done between the open loop and closed loop configuration. For the closed loop configuration, it results that the Miller amplifier (independently on the passive or active low) has a negligible offset until 100 MHz. This result is reasonable because the feedback reduces the output impedance and therefore the susceptibility. Moreover, in this case, by choosing one of the several architectures depicted in the literature, with a high immunity to the EMI injected into the input pin [8–10,13–16], the susceptibility to the output pin can be further reduced, ref. [20]. For the open loop configuration, the behavior is rather unpredictable, but it does not depend on the transistor sizing because a common source stage with the same size and biasing has a different behavior (similar to the one depicted in section III). By removing the RC network, the EMI-induced offset assumes more reasonable values. Therefore, a possible cause of the very large offset at the middle high frequencies can be found by comparing the output node with and without the frequency compensation network. We found that the RC decreases the impedance seen from the drain of the PMOS transistor, making it similar to that of the passive load. In brief, the output node could be in a kind of tri-state. An additional voltage buffer could be useful in this case to reduce the output impedance and therefore the susceptibility. The results obtained in the open loop configuration can be also exemplary to understand the behavior of several circuits, such as digital ones and logic gates. Finally, a general consideration is that the behavior at very high frequencies is always difficult to predict. Neither the measurements are so determining, because of the parasitics of the IC package which complicate the analysis.

9. Conclusions

This paper compares the susceptibility to the interferences coupled to the output pin in several different topologies, starting from the single stage amplifiers up to the whole operational Miller amplifier. The output pin is indeed a critical point of injection and therefore effort and attention must be spent to understand the coupling mechanism. The investigation of single stage amplifiers, such as common source, common gate, common drain and cascode, is easier, but non trivial. The susceptibility can be related, indeed, to the output impedance. Regarding the two stages amplifier, similar considerations can be drawn but also a distinction must be done between the open loop and closed loop configuration. In general, the open loop configuration exhibit a larger and less predictable EMI induced offset, while the amplifier with the feedback loop shows a good immunity up to hundreds of MHz.

Author Contributions: All the authors have contributed substantially to the paper. A.R. has provided the simulation tools and has written the paper; L.C. and Z.K.-V. have written the paper and supervised the work. All authors have read and agreed to the published version of the manuscript.

Funding: This research received no external funding.

Conflicts of Interest: The authors declare no conflict of interest.

Refernces

1. Van De Beek, S.; Leferink, F. Robustness of a TETRA Base Station Receiver Against Intentional EMI. *IEEE Trans. Electromag. Compat.* **2015**, *57*, 461–469. [CrossRef]
2. Dawson, J.F.; Flintoft, I.D.; Kortoci, P.; Dawson, L.; Marvin, A.C.; Robinson, M.P.; Stojilovic, M.; Rubinstein, M.; Menssen, B.; Garbe, H.; et al. A Cost-Efficient System for Detecting an Intentional Electromagnetic Interference (IEMI) attack. In Proceedings of the IEEE International Symposium on Electromagnetic Compatibility, Gothenburg, Sweden, 1–4 September 2014; pp. 1252–1256.
3. *IEEE Draft Recommended*; Practice for Protecting Public Accessible Computer Systems from Intentional EMI; IEEE P1642/D10; IEEE: Piscataway, NJ, USA, 2014; pp. 1–31.
4. Kune, D.F.; Backes, J.; Clark, S.S.; Kramer, D.; Reynolds, M.; Fu, K.; Kim, Y.; Xu, W. Ghost Talk: Mitigating EMI Signal Injection Attacks against Analog Sensors. In Proceedings of the 2013 IEEE Symposium on Security and Privacy, Berkeley, CA, USA, 19–22 May 2013; pp. 145–159.
5. Van De Beek, S.; Leferink, F. Current intentional EMI studies in Europe with a focus on STRUCTURES. In Proceedings of the IEEE International Symposium on Electromagnetic Compatibility Tokyo 2014, Tokyo, Japan, 12–16 May 2014; pp. 402–405.
6. Fiori, F.; Crovetti, P. Prediction of high-power EMI effects in CMOS operational amplifiers. *IEEE Trans. Electromagn. Compat.* **2006**, *48*, 153–160. [CrossRef]
7. Ramdani, M.; Sicard, E.; Boyer, A.; Ben Dhia, S.; Whalen, J.J.; Hubing, T.H.; Coenen, M.; Wada, O. The Electromagnetic Compatibility of Integrated Circuits-Past, Present and Future. *IEEE Trans. Electromagn. Compat.* **2009**, *51*, 78–100. [CrossRef]
8. Crovetti, P.S. Operational amplifier immune to EMI with no baseband performance degradation. *Electr. Lett.* **2010**, *46*, 209–210. [CrossRef]
9. Walravens, C.; Van Winckel, S.; Redoute, J.M.; Steyaert, M. Efficient reduction of electromagnetic interference effects in operational amplifiers. *Electr. Lett.* **2007**, *43*, 84–85. [CrossRef]
10. Richelli, A. CMOS OpAmp resisting to large electromagnetic interferences. *IEEE Trans. Electromagn. Compat.* **2010**, *52*, 1062–1065. [CrossRef]
11. Fiori, F. EMI susceptibility: The achilles' heel of smart power ICs. *IEEE EMC Mag.* **2015**, *4*, 101-105. [CrossRef]
12. Crovetti, P.; Musolino, F. Interference of Spread-Spectrum EMI and Digital Data Links under Narrowband Resonant Coupling. *Electronics* **2020**, *9*, 60. [CrossRef]
13. Richelli, A.; Matig-a, G.; Redoute, J.M. Design of a folded cascode opamp with increased immunity to conducted electromagnetic interference in 0.18 μm CMOS. *Microelectr. Reliab.* **2015**, *55*, 654–661. [CrossRef]
14. Redoute, J.M.; Steyaert, M. EMI-Resistant CMOS Differential Input Stages. *IEEE Trans. Circuits Syst. I Regul. Pap.* **2010**, *57*, 323–331. [CrossRef]
15. Richelli, A. Increasing EMI immunity in novel low-voltage CMOS OpAmps. *IEEE Trans. Electromagn. Compat.* **2010**, *54*, 947–950. [CrossRef]
16. Sbaraini, S.; Richelli, A.; Kovács-Vajna, Z.M. EMI susceptibility in bulk-driven Miller opamp. *Electr. Lett.* **2010**, *46*, 1111–1113. [CrossRef]
17. Redouté, J.-M.; Steyaert, M. An EMI Resisting LIN Driver in 0.35-micron High-Voltage CMOS. *IEEE J. Solid State Circuits* **2007**, *42*, 1574–1582. [CrossRef]
18. Richelli, A.; Delaini, G.; Grassi, M.; Redouté, J.-M. Susceptibility of Operational Amplifiers to Conducted EMI Injected Through the Ground Plane into Their Output Terminal. *IEEE Trans. Reliab.* **2016**, *65*, 1369–1379. [CrossRef]
19. Richelli, A.; Colalongo, L.; Toninelli, L.; Rusu, I.; Redouté, J.-M. Measurements of EMI susceptibility of precision voltage references. In Proceedings of the EMCCompo 2017, St. Petersburg, Russia, 4–8 July 2017.

20. Becchetti, S.; Richelli, A.; Colalongo, L.; Kovács-Vajna, Z.M. A Comprehensive Comparison of EMI Immunity in CMOS Amplifier Topologies. *Electronics* **2019**, *8*, 1181. [CrossRef]
21. Preibisch, J.B.; Duan, X.; Schuster, C. An Efficient Analysis of Power/Ground Planes With Inhomogeneous Substrates Using the Contour Integral Method. *IEEE Trans. Electromagn. Compatib.* **2014**, *56*, 980–989. [CrossRef]
22. Kachout, M.; Bel Hadj Tahar, J.; Choubani, F. Modeling of Microstrip and PCB Traces to Enhance Crosstalk Reduction. In Proceedings of the IEEE SIBIRCON-2010, Listvyanka, Russia, 11–15 July 2010; pp. 594–597.
23. Ricchiuti, V. Evaluation of the Effective Electrical Parameters of a PCB Trace for Accurate Signal Integrity Simulations. In Proceedings of the PIERS 2004, Pisa, Italy, 28–31 March 2004; pp. 77–80.
24. Richelli, A.; Kennedy, S.; Redouté, J.M. An EMI-Resistant Common-Mode Cancelation Differential Input Stage in UMC 180 nm CMOS. *IEEE Trans. Electromagn. Compatib.* **2017**, *59*, 2049–2051. [CrossRef]
25. Redouté, J.-M.; Steyaert, M. *EMC of Analog Integrated Circuits*; Springer: Dordrecht , The Netherlands, 2010.
26. Richelli, A.; Redouté, J.-M. A Methodological Approach to EMI Resistant Analog Integrated Circuit Design. *IEEE EMC Mag.* **2015**, *4*, 92–100.
27. Redouté, J.-M.; Steyaert, M. Current mirror structure insensitive to conducted EMI. *IET Electr. Lett.* **2005**, *41*, 1145–1146. [CrossRef]

Article

Hall-Effect Current Sensors Susceptibility to EMI: Experimental Study

Orazio Aiello

Department of Electronics and Telecommunications (DET), Politecnico di Torino, Corso Duca degli Abruzzi 24, I-10129 Torino, Italy; orazio.aiello@polito.it

Received: 27 September 2019; Accepted: 3 November 2019; Published: 8 November 2019

Abstract: The paper deals with the susceptibility to Electromagnetic Interference (EMI) of Hall-effect current sensors. They are usually employed in power systems because of their galvanic isolation. The EMI robustness of such contactless device was compared with that of resistive current sensing (wired method). To this purpose, a printed circuit board (PCB) was fabricated. EMI tests methods such as Bulk Current Injection (BCI), Transverse-Electromagnetic (TEM) cell and Direct Power injection (DPI) were performed to evaluate the robustness of the Hall-Effect current sensor. EMI-induced failures are highlighted by comparing the different measurements tests and setups.

Keywords: hall-effect current sensors; commercial current sensor; electromagnetic compatibility (EMC); electromagnetic interference (EMI); direct power injection (DPI) test; transverse-electromagnetic (TEM) test; bulk current injection (BCI) test

---

1. Introduction

Current sensing circuits are essential for control and monitoring purposes in switching power supplies. In fact, a current sensor is usually employed in the current-mode loop control of DC-DC converters [1–5]. Among the current sensing techniques, Hall-effect current sensors are sensitive to the magnetic field generated by the current to be detected. This allows keeping the power circuit current flows to be monitored electrically isolated from the sensor [6]. On this basis, Hall-effect sensors are not affected by conducted interference generated by the power system to be monitored [7–10] and are therefore particularly employed in electromagnetically-polluted environments.

In this study, the correctness of Hall-effect current sensor operation in the presence of radiated and conducted Electromagnetic Interference (EMI) was investigated referring to a commercial device. A printed circuit board (PCB) suitable to perform Electromagnetic Compatibility (EMC) test was designed so that a continuous wave (CW) Radio Frequency Interference (RFI) could be superimposed on an operating Hall-effect current sensor according to the respective international standard (such as Bulk Current Injection (BCI) [11], Transverse-Electromagnetic (TEM) cell [12] and Direct Power injection (DPI) [13]).

The paper is organized as follows. In Section 2, the test printed circuit board (PCB) is described. Section 3 focuses on the BCI experimental setup. The results of the respective BCI immunity tests are also reported and discussed. In Section 4, the EMI susceptibility of the Hall-effect is further investigated by TEM cell immunity tests. The DPI tests performed to point out the Hall-effect susceptibility to conducted disturbances are presented in Section 5. Finally, some concluding remarks are drawn in Section 6.

2. Test Board

To compare the EMI robustness of a contactless current sensing approach (that exploits the Hall-effect) and a wired one (based on the voltage drop across a sensing resistance), a test PCB was designed. Whenever the Power MOS is switched-on by a driver, the current to be detected flows through the power trace on the PCB. When the setup aims to test a resistive current sensing under the effect of EMI, the Hall sensor output is floating and the voltage across the sense resistor R_{sense} in series with the Power MOS is properly amplified and processed by a detection block (as sketched in Figure 1a). On the contrary, to evaluate the EMI robustness of the Hall sensor, the resistor R_{sense} is shorted and the output voltage of the Hall sensor is elaborated by a properly-modified detection block (as sketched in Figure 1b).

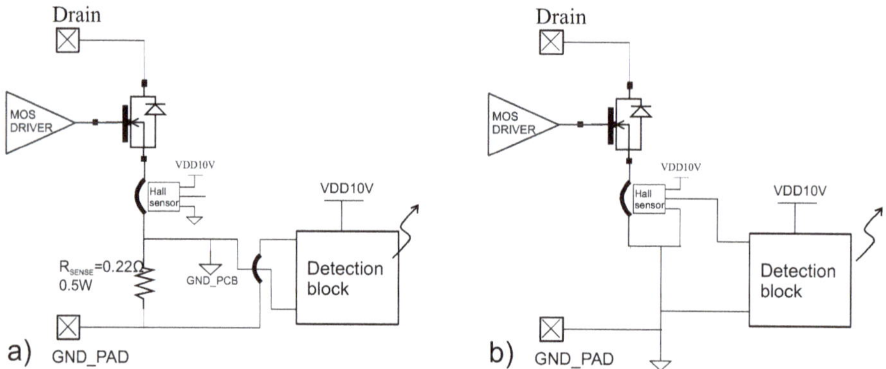

Figure 1. Schematic view of the current sensing setup for: (**a**) resistive approach; and (**b**) the Hall sensor.

A schematic layout representation of the the test PCB that can address both the current sensing methods is reported in Figure 2. A power MOS transistor, which is remotely driven by an optical fiber link, is employed to switch on and off the power circuit. The power supplies to the MOS and to the analog front-end are obtained from a line of the wiring harness and reach the PCB trough an automotive connector.

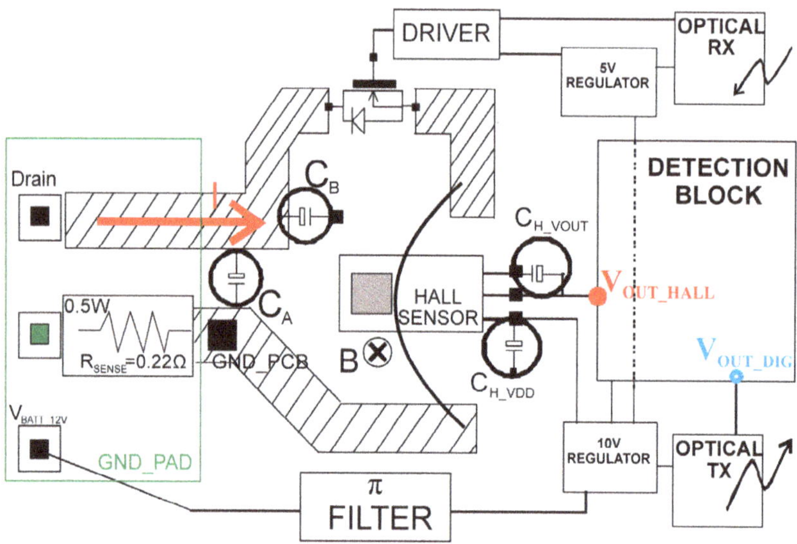

Figure 2. Layout representation of the testing board.

The power supply voltage for the signal acquisition front-end is filtered by a π-type LC and further processed by a 10 V regulator. To reject conducted disturbance in the range of the DPI tests (1–400 MHz), the π filter is composed of a 3 µH inductor and two 100 nF capacitors at its terminals. The regulator provides the power supply to the Hall sensor, to the optical transmitter and to the detection block. A further voltage regulator provides a 5 V supply to the optical receiver and power driver.

The bottom side of the PCB is a ground reference plane named GND_PCB. In Figure 2, the rectangular box on the left indicates a track on the bottom side of the PCB that separates two different ground plane references: GND_PAD and GND_PCB (respectively, inside and outside the rectangular box). This two reference areas are electrically connected by the resistance R_{sense} during resistive current sensing, while, for the Hall-effect based current sensing, the two areas are shorted. Electromagnetic simulations by means of ANSYS Maxwell [14] were performed on the PCB cross section to ensure that the Hall-effect device did not suffer from the magnetic field due to the current in the bottom ground plate. Thus, only the wire placed on the Hall sensor (as in Figure 2) generated the magnetic field due to the current flowing in itself and to be detected by the sensor.

Hall-Effect versus Resistive Setup

Depending on which of the two considered current sensing is investigated (Hall-effect based or resistive one), proper paths are enabled and the respective components are mounted on the PCB to define the signal processing chain. Thus, the Hall sensor output voltage or the voltage drop across R_{sense}, respectively, is properly processed by a detection block, as represented in Figure 2, which contains a non-inverting gain stage and a hysteresis threshold comparator, as reported in Figure 3. Such a hysteresis comparator provides a digital output voltage related to its threshold input voltages (output voltage of the amplifier), respectively, equal to V_{TL} and V_{TH}. Such a comparator drives an optical transmitter so that the comparator output can be remotely monitored.

For resistive current sensing, these hysteresis thresholds correspond to a sensed current, respectively, equal to $I_{\text{min}} = 0.85$ A and $I_{\text{max}} = 0.65$ A, as reported in Figure 4a.

For Hall-effect current sensing, depending on the magnitude of the current flowing through the power circuit, a magnetic induction field is generated. The Hall-effect sensor is sensitive to this magnetic induction field and generates a proportional output voltage showing. The sensitivity is $1\,\frac{\text{mV}}{\text{mT}}$, which roughly corresponds to an output voltage of 10 mV for a current of 1 A flowing in the PCB power circuit. On this basis, the two voltage levels V_{TL} and V_{TH} are due to two correspondent current values equal, respectively, to $I_{\text{min}} = 0.9$ A and $I_{\text{max}} = 2.7$ A, as reported in Figure 4b. Notice that these two current values vary depending on the distance between the Hall magnetic sensitive plate and the wire in which flows the current that generates the sensed magnetic field.

The hysteresis windows in Figure 4a,b are defined to find a failure event in the current sensing induced by the EMI-presence. Such event corresponds to the High-to-Low transition of the hysteresis comparator that, in turn, produces a light signal generated by the optical RX that can be remotely acquired.

Despite of the fact that the hysteresis window amplitude for resistive current sensing is 10× narrower than the one employed with the Hall sensor (as can be highlighted comparing Figure 4a,b), the resistive current method shows a strong robustness to EMI. No failure in its current detection were experienced during any of the EMI tests discussed in this paper. Thus, for the sake of simplicity, the robustness of the resistive current method is not mentioned again so that only the investigations referring to the Hall-effect current sensor are discussed in the following.

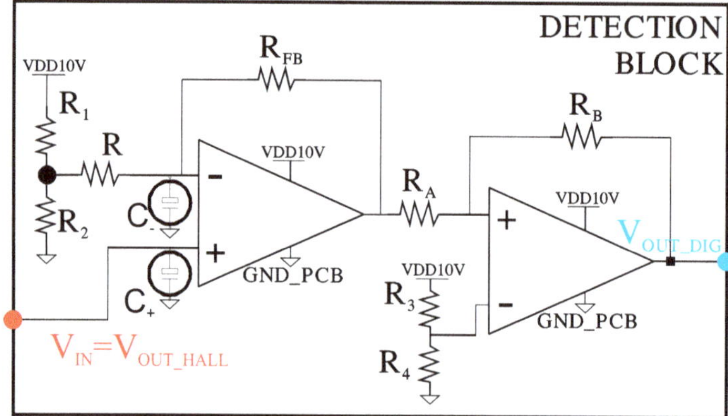

Figure 3. Processing chain of the Hall sensor output voltage.

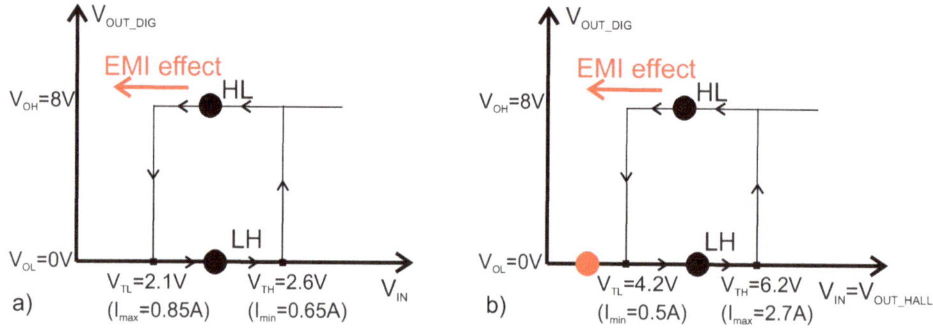

Figure 4. Hysteresis window of the output comparator for: (**a**) resistive current sensing; and (**b**) the Hall-effect sensor.

3. Bulk Current Injection (BCI) Test

To assess the susceptibility of the testing board introduced in Section 2, BCI tests were performed on the overall current detection system in Figure 5 by the setup in Figure 6 [11].

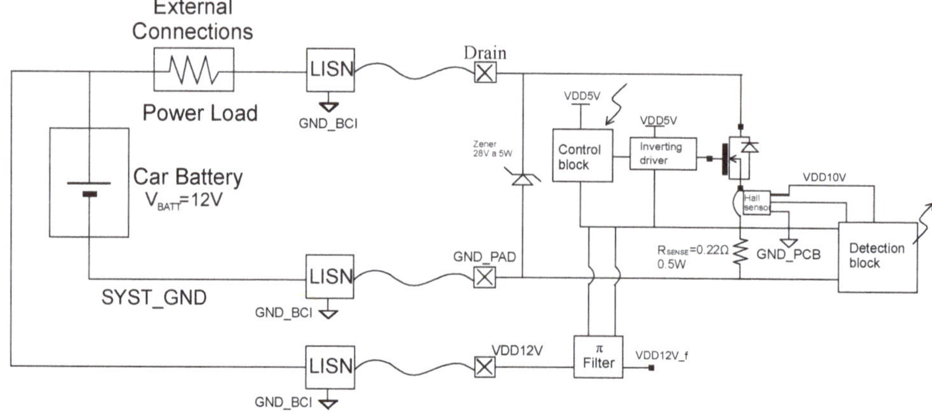

Figure 5. Overall schematic view of the current detection setup.

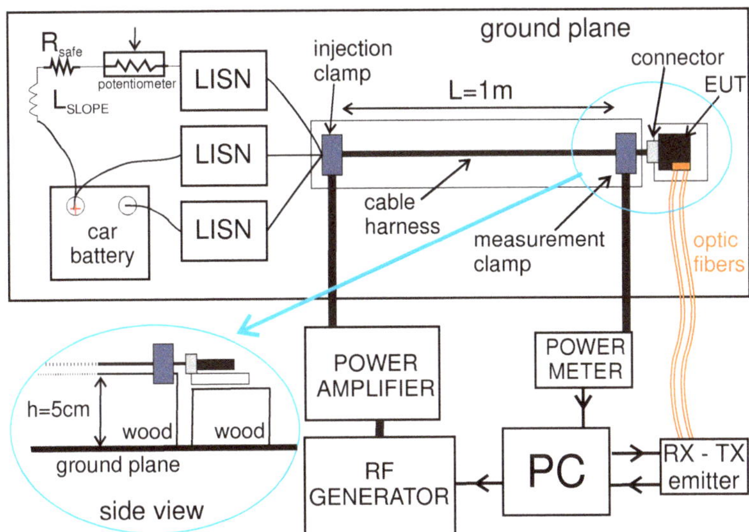

Figure 6. Pictorial representation of BCI test setup [11].

In such a setup, the equipment under test (EUT) is connected to the a variable resistive power load (potentiometer in the figure), to a car battery and to a remote ground reference through a 1 m-long three-wire harness running 5 cm above the system ground plane on a wooden structure (inset in Figure 6). The three wires of the harness are connected to the loads and to the battery through three Line Impedance Stabilization Networks (LISNs in Figures 5 and 6), which provide a 50 Ω RF termination to each line. Every LISN is designed to show an impedance equal to 50 Ω at its RF terminals independently of the load connected at its low-frequency input terminals in the test bandwidth (0–400 MHz).

During the tests, an injection clamp (F-130A-1 [15]) and a measurement clamp (F-51 [16]) by Fischer Custom Communication (FCC) are located along the harness, as shown in Figure 6, to perform BCI tests in compliance with [11]. The distance between the injection and measurement clamps is 75 cm, while the distance between the measurement clamp and the EUT is 4.5 cm. The injection clamp is connected to the output of an 10 W RF power amplifier, whose input terminal is connected to a CW RF signal source, while the measurement clamp is connected to an RF power meter to measure the bulk current injected into the EUT. The power amplifier, the RF source, the RF power meter and the control unit are located out of the test area. The EUT is connected to a two-way optical fiber link in order to drive the power transistor in the EUT and to monitor the current sensor output without perturbing the surrounding electromagnetic environment. A potential malfunction of the current detecting of the EUT during BCI tests is due to common-mode RF current (bulk current) injected into the EUT through its wiring harness.

3.1. BCI Susceptibility Tests

BCI measurements in the bandwidth (10–400 MHz) were performed by a PC-based acquisition system, as represented in Figure 6. For each test frequency, the RFI amplitude is increased until a failure in the DUT operation is experienced or the maximum incident RF power deliverable by the amplifier is reached. In the first case, the failure injected bulk current obtained from the power meter measurement is acquired. Otherwise, no failure value is reported (missing points and respective connecting lines in Figures 7–10).

Referring to the above-mentioned EUT, the commutation of the hysteresis comparator voltage has been considered as a failure criterion and the potentiometer has been configured so that a DC current $I = \frac{I_{\min}+I_{\max}}{2}$ flows in the power circuit. More precisely, at each failure event, the EUT has been configured so that the hysteresis comparator operates at the point HL in Figure 4 with no EMI excitation.

To highlight the susceptibility of the Hall-effect sensors, BCI immunity tests were performed the on different configurations of the EUT introduced in Section 2.

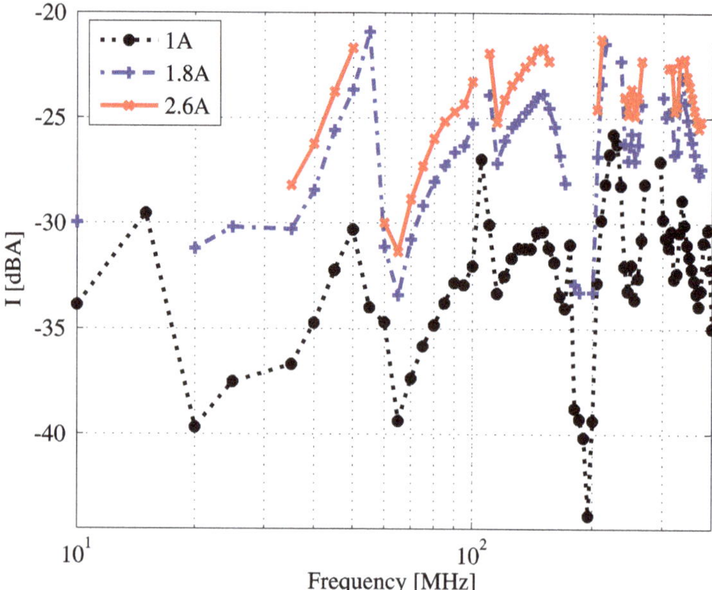

Figure 7. BCI immunity measurement for different current values.

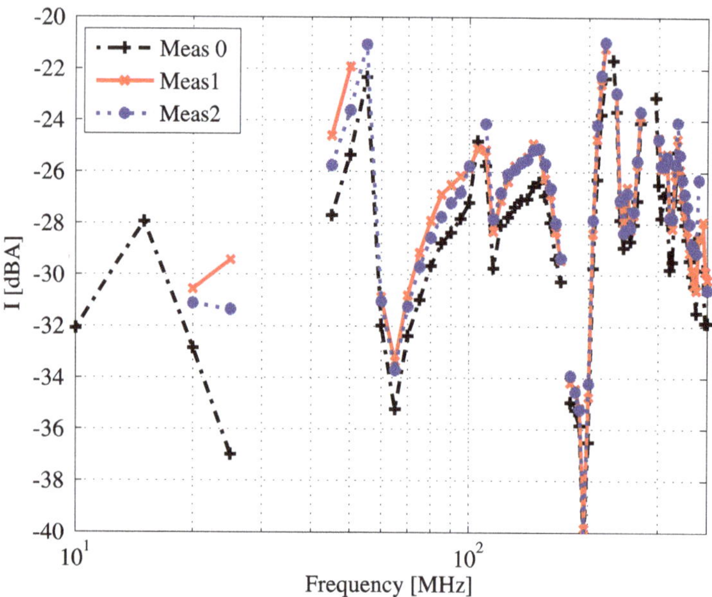

Figure 8. BCI immunity measurement for different caps presence in layout in Figure 2: Meas0 = no caps; Meas1 = C_+, C_-; and Meas2 = C_+, C_-, C_{H_VOUT}, C_{H_VDD}.

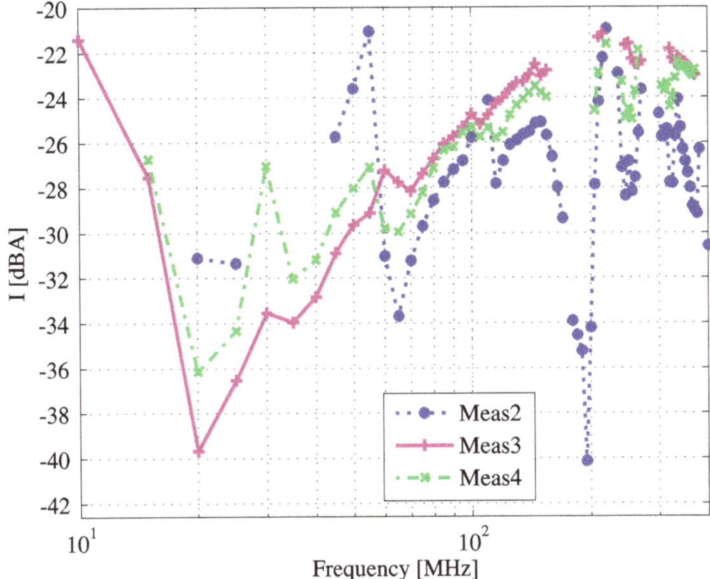

Figure 9. BCI immunity measurement for different caps presence in layout in Figure 2: Meas2 = C_+, C_-, C_{H_VOUT}, C_{H_VDD}; Meas3 = C_+, C_-, C_{H_VOUT}, C_{H_VDD}, C_A; and Meas4 = C_+, C_-, C_{H_VOUT}, C_{H_VDD}, C_B

3.2. BCI Test Results

The susceptibility to EMI of the EUT in Figure 6 was first tested for different values of the DC current flowing through the power line. The results of such tests are reported in Figure 7. It can be observed that the susceptibility level of the EUT scales with the DC current level, depending on the proximity of the threshold value.

Moreover, several BCI measurements were performed adding 100 nF filter capacitors in order to reject the conducted disturbances. They were placed at the terminals of the main blocks (i.e., inputs of the detection block or Hall sensor leads) and to define the influence of the parasitic of the long trace on the PCB.

The encircle capacitances in Figures 2 and 3 point out the most significant filter capacitances added on the PCB in the successive tests described in this section. The susceptibility of the EUT with no filter capacitors on the Hall sensor acquisition front-end is firstly considered and named Meas0 in Figure 8. Then, two capacitors C_+ and C_- were placed at the inputs of the operational amplifier of the detection block, as in Figure 3. The susceptibility tests were repeated and data collected as Meas1 in Figure 8. Another configuration, listed as Meas2 in Figure 8, includes the capacitors C_+ and C_- as in Meas1 but two further capacitors C_{H_VOUT}, C_{H_VDD} at Hall sensor leads, as represented in Figure 2. Comparing the above-mentioned three configurations, the capacitors do not strongly affect the immunity level of the EUT.

Other measurements were performed including other capacitors. The immunity level of the EUT was explored with a capacitor C_A between the drain and source of the power transistor, as reported in Figure 2. At the capacitors already placed in Meas2, such capacitor C_A was added and the respective measurements are named as Meas3 in Figure 9. Then, instead of the capacitor C_A, a capacitor C_B between the drain and the ground plane on the bottom side of the DUT was placed. The respective measurements are reported as Meas4 in Figure 9. Comparing Meas2, Meas3 and Means4 shows that the respective additional capacitors result in a slightly higher immunity for frequencies higher than 600 MHz. Thus, no significant immunity improvement is experienced in any of the above-mentioned configurations.

Other EMI investigations were performed for different Hall sensor inclination with respect to the PCB surface, as represented in Figure 10a. In all cases, the distance beetween the wire and the Hall sensor has not been changed. The experimental results obtained with such configurations are reported in Figure 10b and show a significant dependence of the susceptibility level on the inclination of the sensor.

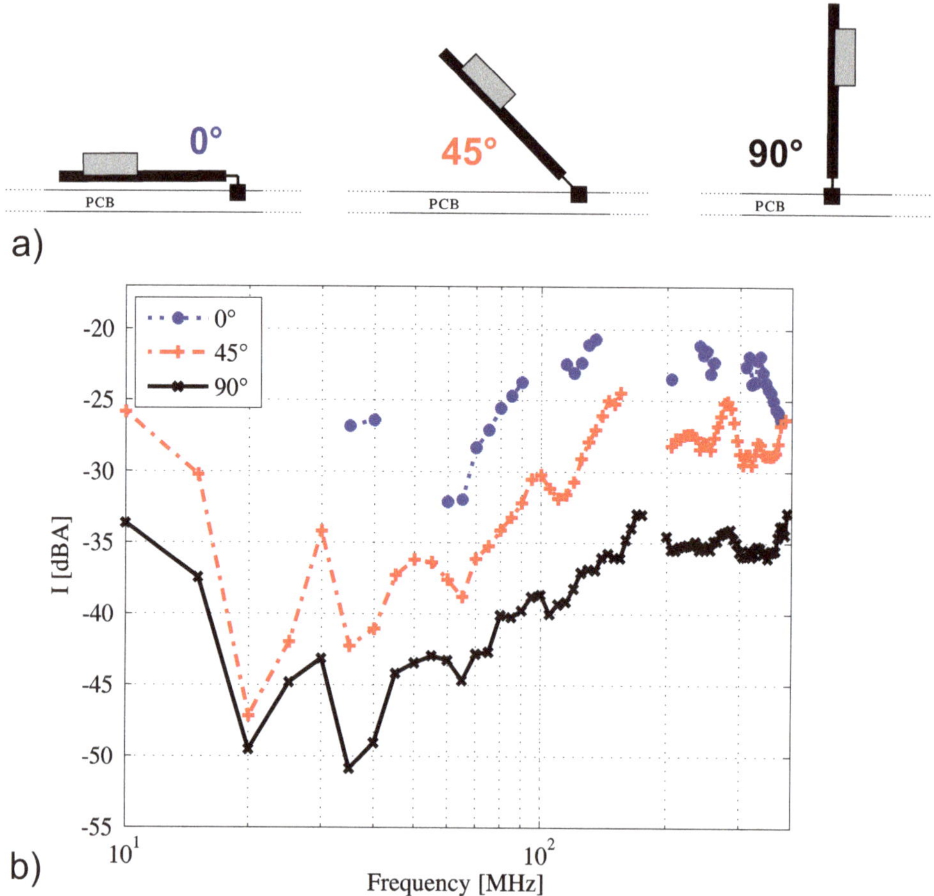

Figure 10. (**a**) Hall sensor inclination considered in BCI tests; and (**b**) respective BCI immunity measurements.

3.3. Discussion

Figure 7 shows how the failure level of the EUT scales with the current to be measured. As a consequence, EMI-induced failures seems to be related to an EMI-induced upset in the Hall-effect sensor output signal path. On the other hand, Figures 8 and 9 show that the susceptibility level of the EUT is scarcely affected by the introduction of filter capacitors at the conditioning amplifier inputs and at the terminals of the Hall-effect current sensors. Therefore, the EUT susceptibility is not related to the detection block or the trace parasitics. On the contrary, the measurements performed for different Hall sensor inclinations (Figure 10b) show that the immunity level is strongly affected by the inclination of the sensor. On this basis, the susceptibility to EMI of the EUT seems to be related to direct electromagnetic field coupling in the Hall-effect sensor body. For this reason, further TEM cell immunity tests were performed.

4. Transverse-Electromagnetic (TEM) Cell Tests

To investigate in further detail the EMI-induced failures in the Hall sensor highlighted by BCI measurements, the susceptibility to radiated EMI of such a sensor was tested in a TEM cell, by placing the sensor in the cell, as sketched in Figure 11. The Hall sensor ground lead was shorted to the TEM cell internal side while the supply voltage (VDD) and the output (V_{OUT}) terminals were AC-shorted to the internal side of the TEM cell. VDD and V_{OUT} leads come out from the TEM cell upper side. Through these terminals, the DC power supply of the sensor was provided and its DC output voltage was measured. To verify the device susceptibility to electric and magnetic fields, the Hall sensor was placed inside the cell TEM in different orientations (A–D in Figure 11).

Figure 12 points out the magnetic and electric field directions in the different setup considered in the following. In Setups A and B, the RF electric field E_x (along x-axis) is parallel to the device orientation. Instead, Setups A and B differ to the device orientation with respect to the RF magnetic field H_y (along y-axis). In Setup A, H_y is parallel to the Hall sensor surface, while, in Setup B, H_y is orthogonal to the device. Similarly, in Setups C and D, the RF electric field E_x is orthogonal while these two setup differ to the RF magnetic field orientation with respect to Hall sensor surface.

4.1. (TEM) Cell Measurement Result

The radiated EMI susceptibility tests highlight that electromagnetic field distribution inside the TEM cell induces a negative offset in the Hall sensor output voltage. In Figure 13, the magnitude of the induced offset voltage in the range from 500 kHz to 1 GHz for an RF incident power applied to the TEM cell equal to 34 dBm. This implies a vertical electric field magnitude of 250 V/m and a transversal magnetic field magnitude of 0.66 A/m where the Hall sensor under test is placed [17–19].

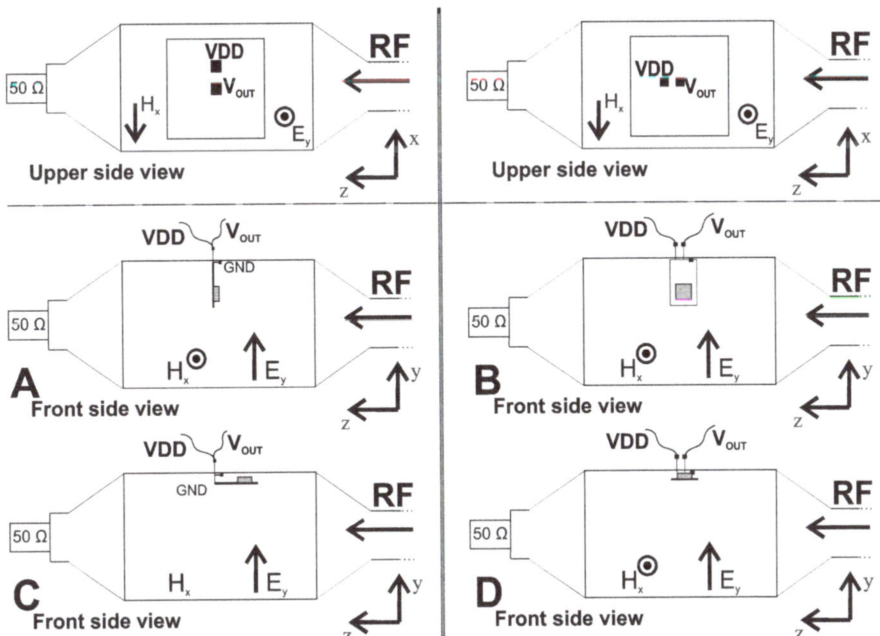

Figure 11. TEM cell testing setup.

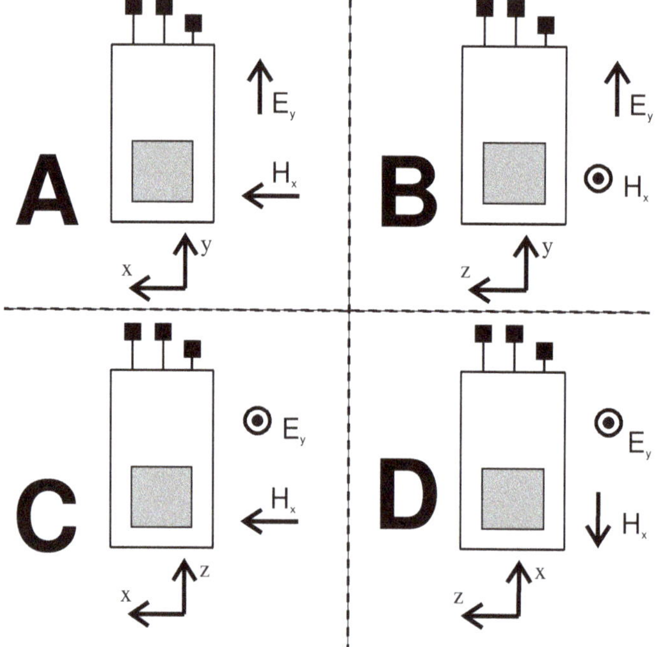

Figure 12. Magnetic and electric field orientation for different Hall sensor setup.

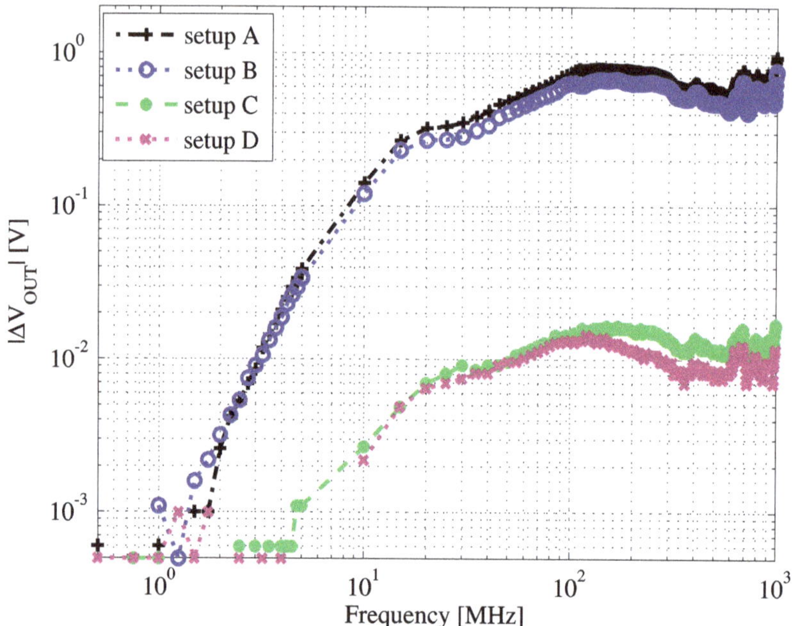

Figure 13. TEM cell immunity measurements at 34 dBm RF power amplitude for different setup.

4.2. Discussion

Figure 13 highlights the similarity between the EMI induced offset voltage measured in Setups A and B. In addition, Setups C and D show similarity in terms of EMI-induced offset. Moreover, Setups A and B are differently coupled to the TEM cell magnetic field and similarly coupled to the electric field, whereas Setups A and C (Setups B and D) are differently coupled in terms of electric field. On this basis, the susceptibility to EMI of the Hall-effect sensor can be related to the electric field parallel to the Hall sensor plate. This could be due to the susceptibility of the sensor either to the electric field directly coupled to the sensing element or to the disturbances connected by leads and metal interconnections on the sensor body.

5. Direct Power Injection (DPI)

As a further EMI investigation, direct power injection tests [13] on the employed Hall sensor were performed. RF power was injected on the supply voltage of the Hall sensor by means of an RF generator, an RF amplifier and a bias tee, as represented in Figure 14. The employed bias tee has a lower bandwidth limit equal to 30 MHz. Therefore, in the following figure, the measurement results are not reliable for a frequency lower that 30 MHz. At the end of the power amplifier a -3 dB attenuator was placed for the safety of the RF amplifier. In fact, if the RF amplifier were accidentally not totally connected (open circuit condition) during the measurement operation, a reflection coefficient $|\Gamma| = 1$ could damage the RF amplifier.

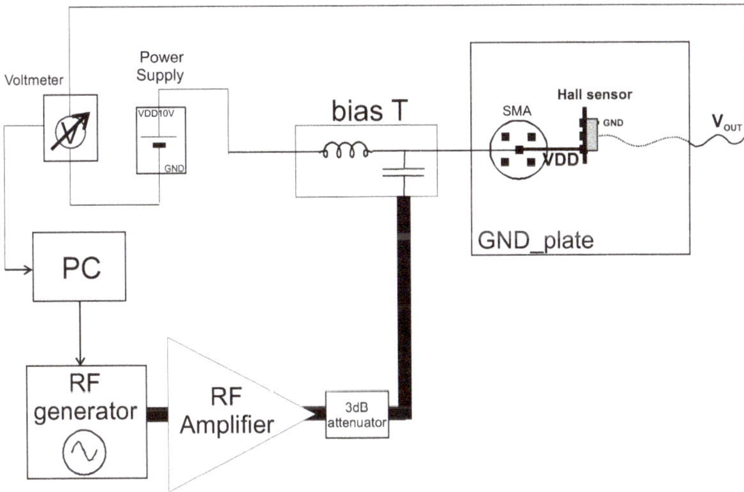

Figure 14. DPI on Hall sensor VDD Test setup [13].

The incident powers provided are given as results of the sum of the incident power of the RF generator (respectively, -50, -40, -30, -20, and -10 dBm), 34 dBm of the RF amplifier and the attenuation due to a -3 dB attenuator. On this basis, the induced offset voltages ΔV_{OUT} due to RF power amplitudes equal -19, -9, 1, 11, and 21 dBm injected on VDD in the range 500 kHz–1 GHz are reported in Figure 15. Similarly, the induced offset voltages ΔV_{OUT} due to frequencies of 100 MHz, 500 MHz and 1 GHz for amplitude in the range -19 dBm to +21 dBm are reported in Figure 16. An upper limit to provided RF power amplitude equal to 21 dBm was chosen to avoid Hall sensor destruction due an excessive power at its terminals. Such an amplitude corresponds to a theoretical maximum voltage on supply voltage VDD lead equal to 17 V.

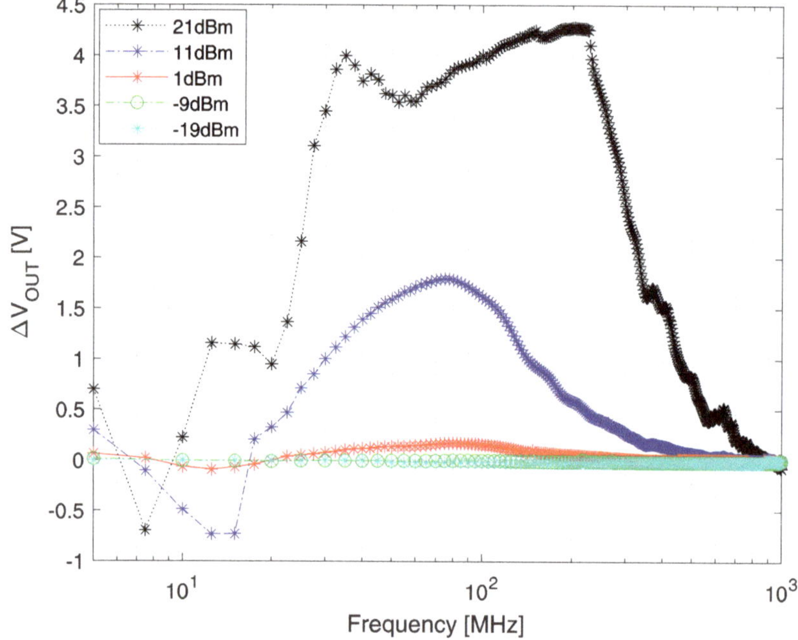

Figure 15. DPI on VDD immunity measurements ($|\Delta V_{OUT}|$ vs. frequency) for different RF power amplitudes.

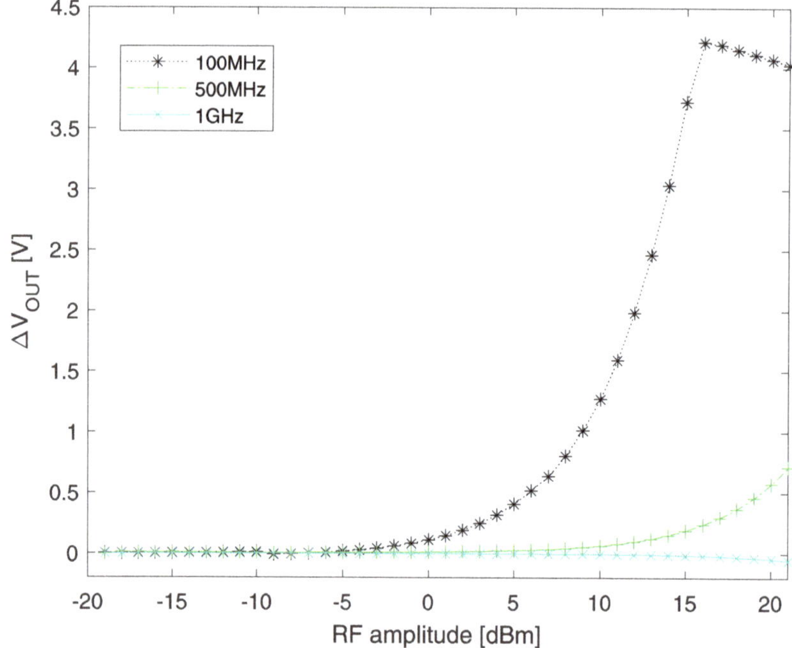

Figure 16. DPI on VDD immunity measurements ($|\Delta V_{OUT}|$ vs. RF power amplitude) for different frequencies.

Different signs of the induced offset are highlighted in Figures 15 and 16. These phenomena seem to be due to different nonlinear mechanisms of active elements in the device. Similar measurement results are shown even in the EMI superimposed on the output terminal of the Hall sensor. Although even the DPI tests show the susceptibility to EMI of the investigated Hall sensor, no precise reasons for such failure can be inferred.

6. Conclusions

The susceptibility to EMI of a Hall-effect sensor employed in current monitoring was investigated firstly referring to BCI test. Hall-effect current sensors (contactless) are dramatically affected by the EMI presence, contrary to the resistive (wired) current sensing method. BCI tests were performed in difference configurations, in addition to TEM cell and DPI tests. The measurements results show that a Hall-effect sensor can be strongly affected by the presence of EMI. In particular, the TEM cell tests highlighted how the operations of the specific Hall-effect sensor considered in this study are affected by an RF electric field excitation parallel to the surface of the sensing element. As the physical mechanism of the Hall-effect is not affected by a CW RFI [7], the failures of the Hall sensor are likely related to the ICs that interface the Hall sensor and process its signal. Although many investigations are present in the literature regarding amplifier and monitoring ICs [20–22], a precise understanding of the Hall sensor failure causes always relies on a complete knowledge of the ICs that interface with the Hall plates. This implies that the EMI-induced effects have to be taken into account from the first phases of the Hall-effect sensor design.

Funding: This research received no external funding.

Conflicts of Interest: The author declares no conflict of interest.

References

1. Tong, Q.; Chen, C.; Zhang, Q.; Zou, X. A Sensorless Predictive Current Controlled Boost Converter by Using an EKF with Load Variation Effect Elimination Function. *Sensors* **2015**, *15*, 9986–10003. [CrossRef] [PubMed]
2. Min, R.; Chen, C.; Zhang, X.; Zou, X.; Tong, Q.; Zhang, Q. An Optimal Current Observer for Predictive Current Controlled Buck DC-DC Converters. *Sensors* **2014**, *14*, 8851–8868. [CrossRef] [PubMed]
3. Aiello, O.; Fiori, F. A new mirroring circuit for power MOS current sensing highly immune to EMI. *Sensor* **2013**, *13*, 1856–1871. [CrossRef] [PubMed]
4. Aiello, O.; Fiori, F. Current sensing circuit for DC-DC converters based on the miller effect. In Proceedings of the 2013 International Conference on Applied Electronics (AE), Pilsen, Czech Republic, 10–11 September 2013; pp. 1–4.
5. Huang, K.; Liu, Z.; Zhu, F.; Zheng, Z.; Cheng, Y. Evaluation Scheme for EMI of Train Body Voltage Fluctuation on the BCU Speed Sensor Measurement. *IEEE Trans. Instrum. Meas.* **2017**, *66*, 1046–1057. [CrossRef]
6. Popovic, R.S. *Hall Effect Devices*, 2nd ed.; Institute of Physics Publishing: Bristol, UK; Philadelphia, PA, USA, 2004.
7. Aiello, O.; Fiori, F. A New MagFET-Based Integrated Current Sensor Highly Immune to EMI. *Elsevier Microelectron. Reliab.* **2013**, *53*, 573–581 . [CrossRef]
8. Aiello, O.; Crovetti, P.; Fiori, F. Investigation on the susceptibility of hall-effect current sensors to EMI. In Proceedings of the 10th International Symposium on Electromagnetic Compatibility, York, UK, 26–30 September 2011; pp. 1–4.
9. Satav, S.M.; Agarwal, V. Design and Development of a Low-Cost Digital Magnetic Field Meter With Wide Dynamic Range for EMC Precompliance Measurements and Other Applications. *IEEE Trans. Instrum. Meas.* **2009**, *58*, 2837–2846. [CrossRef]
10. Dalessandro, L.; Karrer, N.; Kolar, J.W. High-Performance Planar Isolated Current Sensor for Power Electronics Applications. *IEEE Trans. Power Electron.* **2007**, *22*, 1682–1692. [CrossRef]
11. International Standard ISO 11452-4:2005. Road Vehicles–Component Test Method for Electrical Disturbances from Narrowband Radiated Electromagnetic Energy—Part 4: Bulk Current Injection (BCI). Available online: https://www.iso.org/standard/37414.html (accessed on 6 November 2019).

12. International Standard IS0 11452-3:2016. Road Vehicles—Electrical Disturbances by Narrowband Radiated Electromagnetic Energy, Component Test Method Part 3: Transverse Electromagnetic Mode (TEM) Cell. Available online: https://www.iso.org/standard/66829.html (accessed on 6 November 2019).
13. IEC 623132-4:2006. Integrated Circuits, Measurement of Electromagnetic Immunity—Part 4: Direct RF Power Injection Method. Available online: https://webstore.iec.ch/publication/6510 (accessed on 6 November 2019).
14. Available online: https://www.ansys.com/products/electronics/ansys-maxwell (accessed on 6 November 2019).
15. Fischer Custom Communication (FCC). *F-130A-1 Injection Current Probe Characterization: Fischer Custom Communication (FCC)*; FCC: Torrance, CA, USA, 2004.
16. Fischer Custom Communication (FCC). *F-51 Monitor Current Probe Characterization: Fischer Custom Communication (FCC)*; FCC: Torrance, CA, USA, 2004.
17. Fiori, F.; Musolino, F. Investigation on the effectiveness of the IC susceptibility TEM cell method. *IEEE Trans. EMC* **2004**, *46*, 110–114. [CrossRef]
18. Fiori, F.; Musolino, F. Measurement of integrated circuit conducted emissions by using a transverse electromagnetic mode (TEM) cell. *IEEE Trans. EMC* **2001**, *43*, 622–628. [CrossRef]
19. Spiegel, R.J.; Joines, W.T.; Blackman, C.F.; Wood, A.W. A method for calculating electric and magnetic fields in TEM cell at ELF. *IEEE Trans. Electromagn. Compat.* **1987**, *EMC-29*, 265–272. [CrossRef]
20. Redoute, J.M.; Steyaert, M. *EMC of Analog ICS*; Springer: Berlin, Germany, 2010.
21. Richelli, A.; Matiga, G.; Redoute, J.M. Design of a Folded Cascode Opamp with Increased Immunity to Conducted Electromagnetic Interference in 0.18 um. *Elsevier Microelectron. Reliab.* **2015**, *55*, 654–661. [CrossRef]
22. Aiello, O.; Fiori, F. On the Susceptibility of Embedded Thermal Shutdown Circuit to Radio Frequency Interference. *IEEE Trans. EMC* **2012**, *54*, 405–412. [CrossRef]

Article

A Novel Meander Split Power/Ground Plane Reducing Crosstalk of Traces Crossing Over

Jung-Han Lee [1,2]

1 Department of Electronic Engineering, Sogang University, Seoul 04107, Korea; leejunghan@kiost.ac.kr
2 Marine Security and Safety Research Center, Korea Institute of Ocean Science & Technology, Busan 49111, Korea

Received: 29 August 2019; Accepted: 16 September 2019; Published: 17 September 2019

Abstract: In this paper, a novel meander split power/ground plane is proposed for reducing crosstalk between parallel lines crossing over it. The working mechanism of the meander split scheme is investigated by simulations and measurements. The LC equivalent circuit and transmission line model are developed for modeling interactions between the meander split and the signal lines. The proposed meander structure enhances electromagnetic coupling between split planes. The capacitive coupling across the split ensures signal integrity and magnetic coupling between adjacent finger shaped structures suppresses lateral wave propagation along the split gap, which in turn helps suppress the crosstalk. The effectiveness of the meander split remains valid over very wide frequency ranges (up to 9 GHz). Experimental results show that the proposed structure improves the signal quality and reduces the near/far end crosstalk over 30 dB and 50% in the frequency domain and time domain, respectively.

Keywords: meander split; power/ground plane; crosstalk; signal integrity; equivalent circuit; capacitive and magnetic coupling

---

1. Introduction

In high-speed electronic systems, the power and ground planes play important roles as a reservoir in supplying power to components and as a voltage reference on printed circuit boards. To accommodate the rapidly switching components and their demand for current, an ideal power supply should have very low impedances, which necessitates the use of power and ground planes. However, the plane pair effectively forms a parallel plate waveguide, which can hold persistent ringing noises generated by routed traces and vias to and from components on the circuit board. To reduce the noise coupling due to power planes and provide different power supply voltages, slotted or split plane types are frequently used for the integrated circuits or modules [1–3] occupying the same printed circuit boards (PCBs) [4,5]. However, power/ground partitioning generates undesired electromagnetic effects such as signal integrity degradation, electromagnetic interference (EMI) and crosstalk when signal lines cross over the split gaps [6–10]. When two parallel line traces cross over slots or splits in the planes, the crosstalk level between the traces becomes high even for large clearances [11].

Commonly used methods to reduce crosstalk are placing via fences [12], guard traces [13], serpentine guard traces [14], stubs [15], or resonators [16] between the two signal lines. Recently, a method of coating signal lines with graphene-paraffin has also been studied [17]. In most of the approaches, efforts are made to decrease the crosstalk levels by inserting additional structures between the conventional transmission lines. Defective microstrip line structures [18] and stub-alternated microstrip lines [19] have been used for the reduction of crosstalk. With these methods, the complexity of the PCBs is increased due to the additional structures.

Attaching stitching capacitors [11,20] or inter-digital capacitors [21] between the split gap under the signal lines reduces the crosstalk and provides return current paths at high frequency while maintaining distinct dc levels of each region. However, this requires additional processes or costs and sometimes it is hard to make room for the stitching capacitor right below the signal trace. Furthermore, the effectiveness of these approaches is limited in that the equivalent series inductance of the capacitors dominates the impedance of a decoupling capacitor at higher frequencies [22]. Another commonly used approach is the addition a low-Q inductor or a thin inductive trace or stubs [23,24] on split power planes. However, this remedy cannot isolate wideband switching noises generated by each functional block on the same PCB, and cannot accommodate different power supply voltages.

Recently, various shapes of defective ground structures (DGS) such as the "U", "V", "H", rectangular, square, circular, ring and dumbbell shape [25–28] have been investigated to design the wideband filter without adding any complexity to the original structure. Some complex shapes have also been studied, which include meander lines [29]. All of these studies focus on the design of the filter using DGS, and it is necessary to study DGS for their crosstalk reduction effect. Recently, several studies [30,31] have investigated the reduction of crosstalk using rectangular or dumbbell DGS shapes on the ground plane, finding that the reduction effectiveness is 20 dB over a frequency range of 5 GHz.

In this paper, a novel meandered DGS embedded on a split plane is proposed and investigated from the view point of signal transmission and crosstalk reduction. The equivalent circuit model based on slot-coupled cavity equivalent circuit and transmission line theory is presented to describe the behavior of the meandered DGS split gap. The meandered structure enhances capacitive coupling across the split planes, which helps signal transmission of the line traces over a split gap. The structure suppresses lateral wave propagation along the slot-line formed by the split gap by the destructive coupling of the magnetic fields of meandering currents on the adjacent slot line sections. In other words, lateral waves excited along the split gap become evanescent, which helps the return current on the power/ground plane be localized. The crosstalk reduction effectiveness of the split plane with meandered DGS holds 30 dB over a wide frequency range up to 9 GHz, and the crosstalk levels are reduced to over 50% of a simple straight split plane such as a rectangular or dumbbell one [30,31], which is verified by measurements.

2. The Meander DGS on Split Power/Ground Plane

2.1. Advantage and Application of the Meander DGS

In general, the shape, size, and periodicity of the DGS affects the frequency characteristics. A periodic DGS can design well-defined stopbands and passbands and achieve the performance such as high-order filters. DGS can be used in many applications in microwave printed circuits such as filters, amplifiers, oscillators, directional couplers, antennas, and multiplayer stack-up PCBs.

The DGS disturbs the current distribution of the ground plane and changes the characteristics of a transmission line crossing over the DGS [32]. The main disadvantage of the DGS is that they break the return current path and cause spurious radiations in the circuits. The DGS will change the impedance of the ground plane, and lead to spurious radiations. However, our proposed meander DGS has the advantage of providing the return current path by enhancing the coupling between adjacent meander lines. Thus, it can be applied as an excellent method to solve EMI problems such as signal integrity, radiation, and crosstalk.

2.2. Design Methodolgy of the Meander DGS

The meander DGS-like asymmetric inter-digital finger in Figure 1c was used in the split ground plane and optimized for the desired frequency of operation. Due to the structure of the meander split on the ground plane, the vertical length l_{tot} and width w_1, w_2 of the meander split dimension were selected according to the design rule. Within the given criteria ($l_{tot} = 2 \times w_1 + 2 \times l_2 + l_3$), the vertical length l_3 and width w_2 of meander split gap contribute to the mutual inductance and capacitive coupling between the meander gaps, respectively, in Figure 1c. The design parameters related to this electromagnetic coupling are independent of the position of the signal line w_L. That is, the w_L does not necessarily have to be located right above the center of l_1 or w_2.

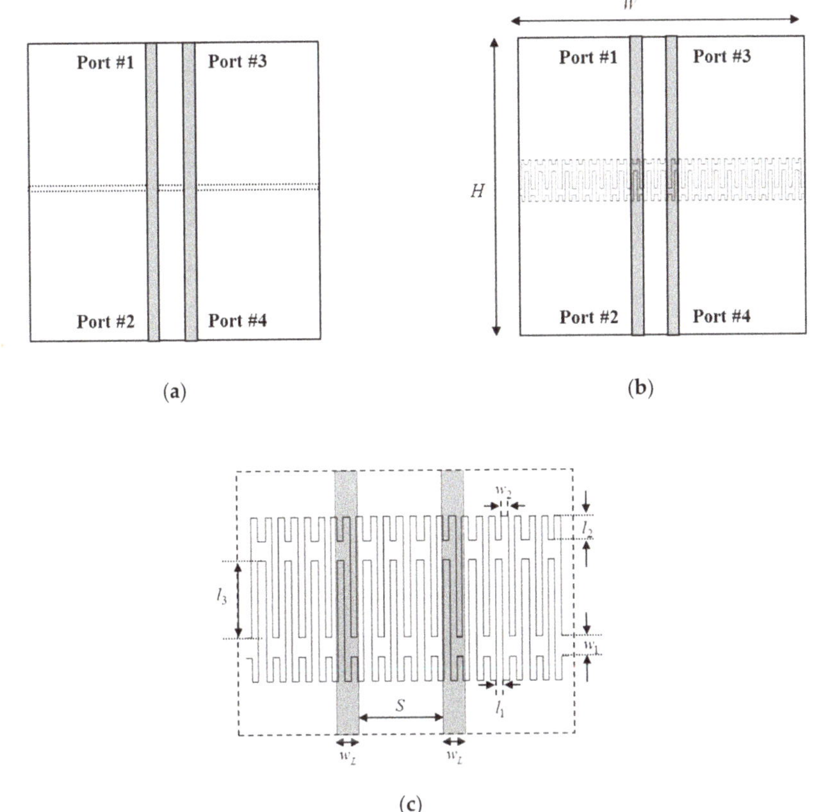

Figure 1. Printed circuit boards with two kinds of split gap adopted. The dotted line is on the bottom side. (**a**) Two signal lines cross over a straight split gap on the ground plane. (**b**) A meander DGS split gap. (**c**) Detailed structure.

The main design method was to select the vertical length l_3 of the meander gap associated with the magnetic coupling and width w_2 of the meander gap associated with the capacitive coupling effect for defining the stopband and passband frequency characteristics. The equivalent circuit of the meander DGS and associated parameters were extracted by the LC equivalent circuit model and transmission line model. The periodic placement (N) of the meander DGS enhances the frequency characteristics of the equivalent circuit and improves the crosstalk reduction effect on broadband frequencies.

Figure 1a,b shows two PCBs containing parallel line traces (W = 100 mm, H = 100 mm, S = 10 mm, w_L = 1.69 mm) on one side. On the bottom sides of the boards, a simple straight split ground plane, like a rectangular DGS [30,31], and a proposed meandered DGS (w_2 = 0.2 mm, l_1 = 0.25 mm, w_1 = 2 mm, l_3 = 20 mm, l_2 = 2 mm) embedded split ground plane are formed, respectively. The dielectric material used for the PCBs is FR4 whose relative permittivity is 4.2, and the thickness of the boards is 1.0 mm. The width of the split gaps of the two boards is 2 mm. There are a total of 110 (N) meandering cells in the split in Figure 1c, which are not drawn to real scale. The characteristic impedances of the signal traces above the solid ground plane are set to be 50 Ω to eliminate reflections from the coaxial cables, which connect the board to a network analyzer.

We measured the scattering parameters of the two samples with a complete homogeneous ground plane in Figure 2. As shown in the figure, the reflections occur due to the split gaps, and the scattering parameters show resonant behavior. Compared with the complete homogeneous ground plane, the two split ground planes degrade the signal integrity and increase crosstalk. However, the meander split improves the signal integrity and decreases 30 dB of crosstalk up to 9 GHz more than the straight split ground plane, like a rectangular DGS [30,31]. This difference comes from the enhanced coupling due to the meander structure on the split plane.

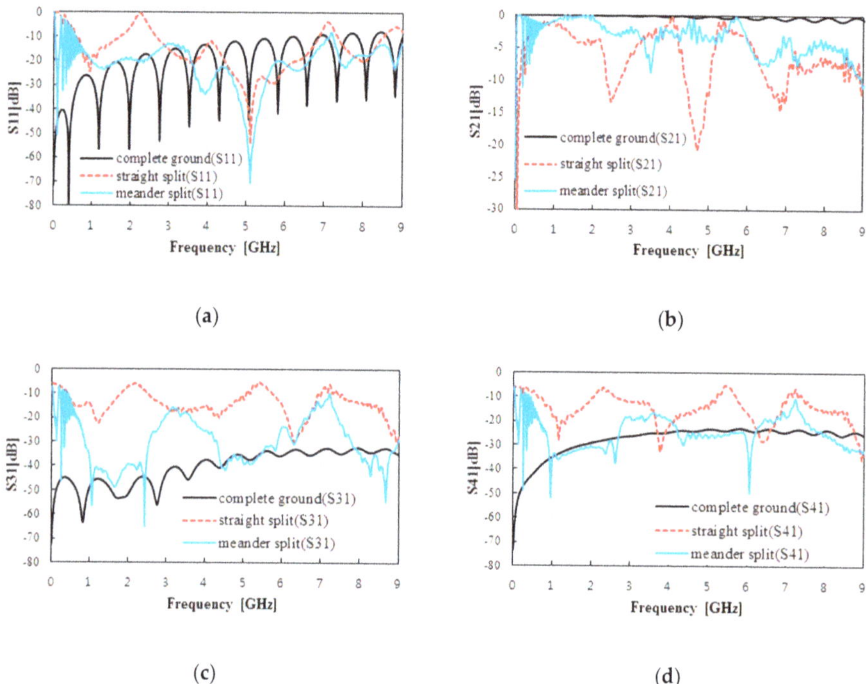

Figure 2. Measured reflection, transmission and crosstalk of the signal line with a complete homogeneous ground, straight and meander split power or ground plane. (**a**) S_{11}. (**b**) S_{21}. (**c**) S_{31} (near-end crosstalk). (**d**) S_{41} (far-end crosstalk).

2.3. Equivalent Circuit of the Meander DGS Split Model

The DGS slot can be modeled in parallel with a capacitor and inductor [32]. Based on the equivalent slot circuit, the LC equivalent circuit of the bottom layer where the ground is a meander gap can be derived in Figure 3. In terms of return current path on the bottom layer, the capacitive and mutual inductance effects exit between the different meanders. The return current paths by the capacitive and mutual inductance effects are represented by the Type 1 (p), Type 2, 3 (m) and Type 4 (n), respectively. The p is the fraction of the return current not involved in the coupling (Type 1), m is the fraction involved in coupling one way only (Type 2 and 3) and n is the fraction involved in coupling one way only (Type 4). The design parameters such as coupling fractions, and the L and C values of the meander structure in Figure 3 were extracted based on the slot-coupled cavity equivalent circuit [33] to define the passband and stopband frequency characteristics.

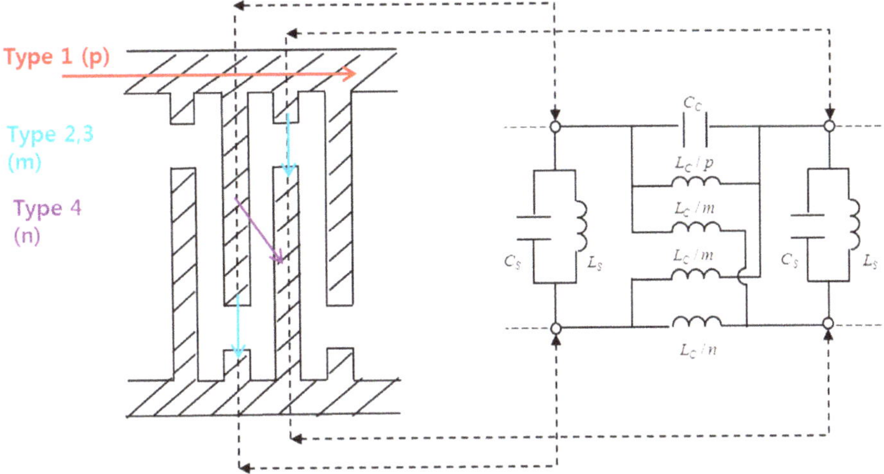

Figure 3. Equivalent circuit of the meander gap.

Alternatively, this enhanced coupling phenomenon can be explained using slot line modes excited in the split. Based on the previous analysis and the scheme of the meander structure, the transmission line equivalent circuit model of the meander split gap can be derived in Figure 4. The meander gap away from the signal line is modeled simply by a series of periodic transmission lines with alternately varying characteristic impedances Z_1, Z_2, Z_3, effective dielectric constants εe_1, εe_2, εe_3 and mutual inductances L_m. To observe the vertical length l_3 and width w_1 of the split gap affecting the crosstalk reduction, we performed a parameter sweep using CST Microwave studio. Due to the structure of the meander split on the ground plane, the vertical length l_{tot} of the meander split cannot be infinitely long. Within the given criteria ($l_{tot} = 2 \times w_1 + 2 \times l_2 + l_3$), the vertical length l_3 of the meander split gap contributes to the mutual inductance. The width w_1 of the meander split gap contributes to the capacitive coupling. We calculated the characteristic impedances and the effective dielectric constants of the slot line as l_3 and w_1 changes (Table 1) by using the conformal mapping technique [34]. By varying the split gap widths (w_1) and length (l_3), the frequency characteristics are designed for passband and stopband of the crosstalk at least (Figure 5).

Table 1. Characteristic impedances and relative dielectric constants of the meander slot line.

l_{tot}(mm)	l_1(mm)	l_2(mm)	l_3(mm)	w_1(mm)	w_2(mm)	w_3(mm)	$Z_1(\Omega)$	$Z_2(\Omega)$	$Z_3(\Omega)$	ε_{e1}	ε_{e2}	ε_{e3}
35	0.25	0.4	5	14.6	0.2	0.2	168	75	75	1.1	1.3	1.3
35	0.25	0.4	15	9.6	0.2	0.2	149	75	75	1.1	1.4	1.4
35	0.25	0.4	30	2.1	0.2	0.2	106	75	75	1.1	1.4	1.4

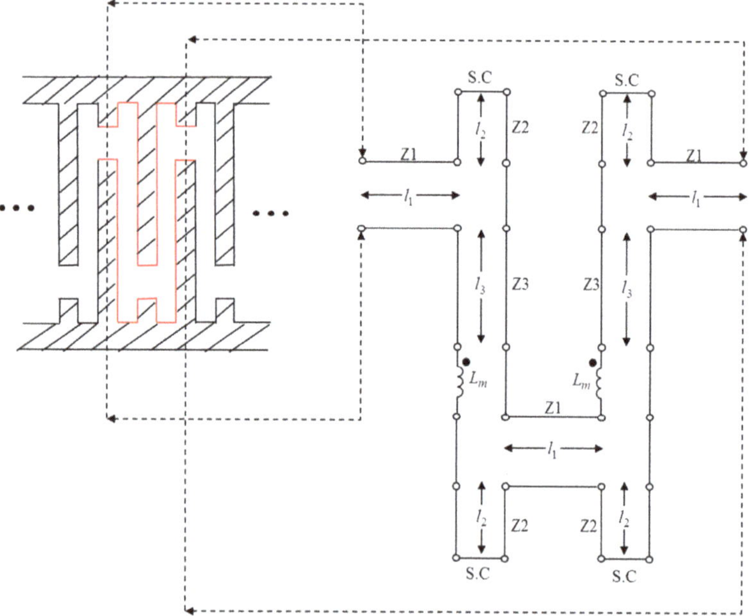

Figure 4. Transmission line equivalent circuit of the meander gap.

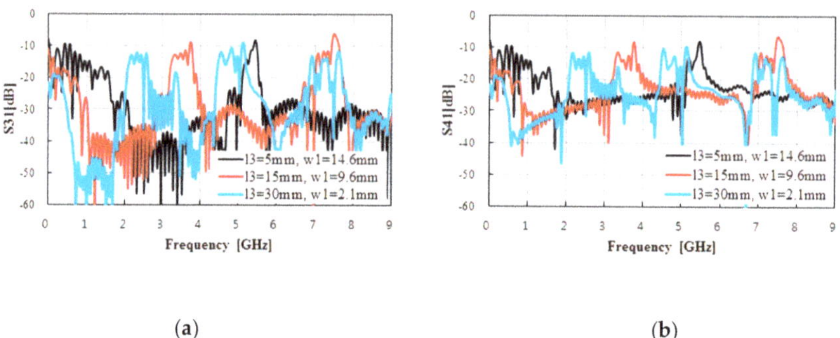

(**a**) (**b**)

Figure 5. S_{31} (near-end crosstalk) and S_{41} (far-end crosstalk) of the meander split plane with length l_3 and w_1 changed. (**a**) S_{31}; (**b**) S_{41}.

Based on the equivalent circuit of the meander gap, the crosstalk equivalent circuit of Figure 6 can be derived for the meander split on the ground plane. The split gap can be modeled by a slot line with a transformer, which is excited by the signal line above [11]. If the spatial period of the meander structure is small enough, the signal line crosses over the split gap in a number of points, which can be modeled by transformers. It shows that if a signal line crosses over the meander split gap on the ground plane, it excites a transverse electromagnetic (TEM) mode in the mender split gap, which has a structure similar to the slot lines. The excited modes propagate in either direction and are scattered at the corners of the meander line, which impede the propagating modes. The split gap right under the signal line can be regarded as a coupled line structure, of which coupling strength can be changed by varying the longitudinal length dimension of the split.

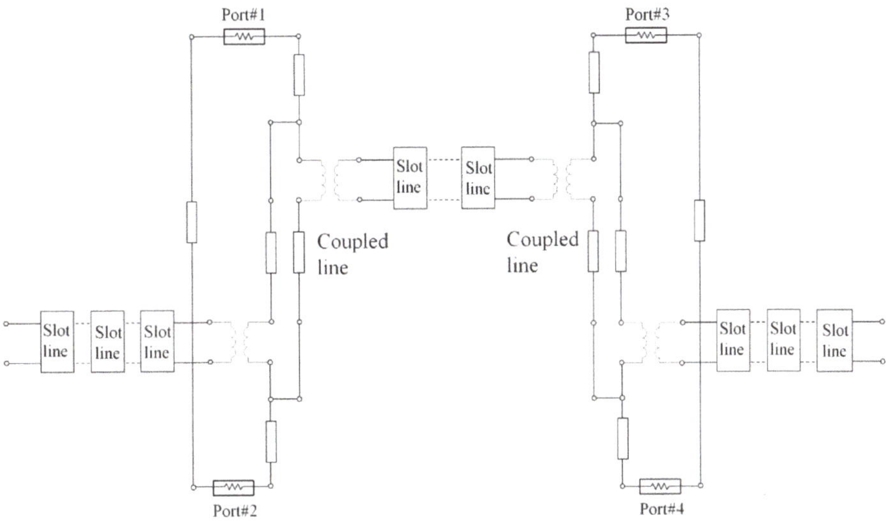

Figure 6. Crosstalk equivalent circuit of the meander split on the ground plane.

3. Results

3.1. Simulation Results of the Equivalent Circuit

Figure 7 shows the effectiveness of the equivalent circuit of Figures 4 and 5. The crosstalk behaviors of the circuit model simulated using Agilent ADS and measurement show good overall agreement about the peak and zero. At low frequencies below about 3 GHz, the LC equivalent circuit model of Figure 4 was closer to the measured result than the transmission line equivalent model of Figure 5. At high frequencies, the transmission line model was close to the measured result. The mutual inductance effect is proportional to the vertical length of the meander split and their relationship is calculated by the linear interpolation from the simulation and measured result. The crosstalk simulation result of the circuit model is slightly overestimated compared to the measured result because of the interpolation residual error of the mutual inductance value.

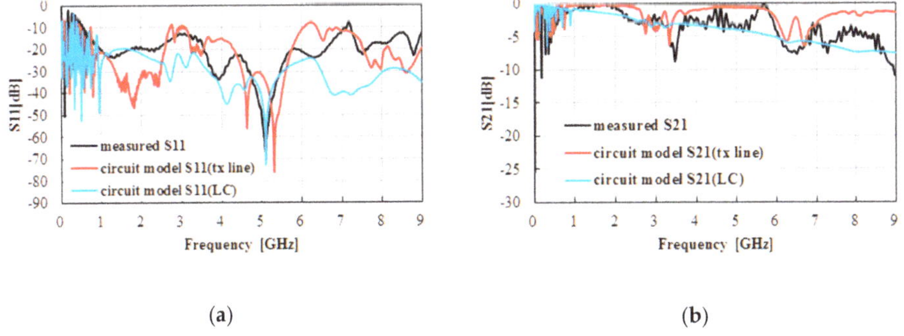

Figure 7. *Cont.*

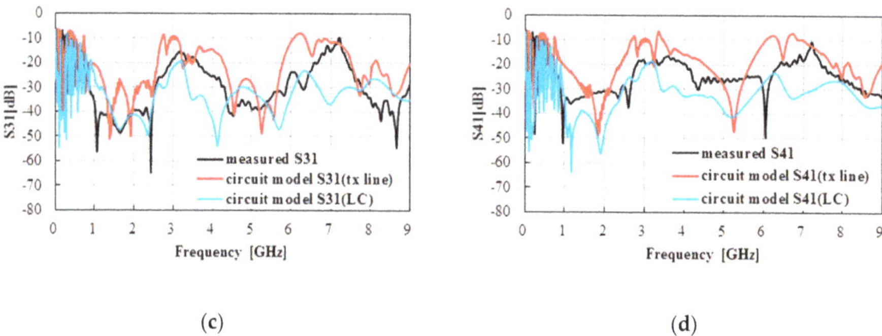

Figure 7. Scattering parameters of the simulation (circuit model) and measured result. (**a**) S_{11}; (**b**) S_{21}; (**c**) S_{31}; (**d**) S_{41}.

3.2. Reduction of Crosstalk According to the Vertical Length and Width of the Meander Split Gap

The effect of reducing the crosstalk is proportional to the vertical length l_3 and l_2 and inversely proportional to the split gap width w_2. We showed the effectiveness of those design parameters using CST Microwave studio. We set the discrete ports at the start and end point of the transmission line equivalent circuit of the meander slot line in Figure 4. Figure 8 shows the more narrowly split gap width w_2 that increases capacitive coupling and the longer length of l_3 and l_2 that causes destructive magnetic coupling with a shortened circuit line and less lateral wave propagation. The shortened circuit transmission lines (l_2) have reactance that impedes lateral wave propagation (Figure 8c). By increasing the number of cells (N) of the meander slot line, as well as varying the meander line length and split gap widths, the excited slot line mode is suppressed or attenuated (Figure 8d).

Therefore, the reduction of crosstalk at the lower and upper frequency has improved, which is desirable for digital signal transmission. The meander split gaps away from the signal line attenuate the scattered waves entering the slot line. Thanks to the evanescent mode, the periodic resonant behavior disappears, which effectively helps the signal transmission and reduces the crosstalk.

These differences come from the enhanced electromagnetic coupling between neighboring slot line sections of split planes. The meander split structure increases capacitive coupling due to the narrow split gap width, and increased length of the interaction causes destructive magnetic coupling due to the oppositely directed currents on nearby slot line sections. The capacitive coupling decreases impedance across the split gap, and the magnetic coupling increases impedances along the split. With the increased impedance, the return current cannot spread perpendicularly to the signal line. The effect causes the return current to be formed near the signal line, which helps the return current flow across the split gap without detouring. The localized return current improves signal integrity of the line traces formed over the split more effectively than the straight split.

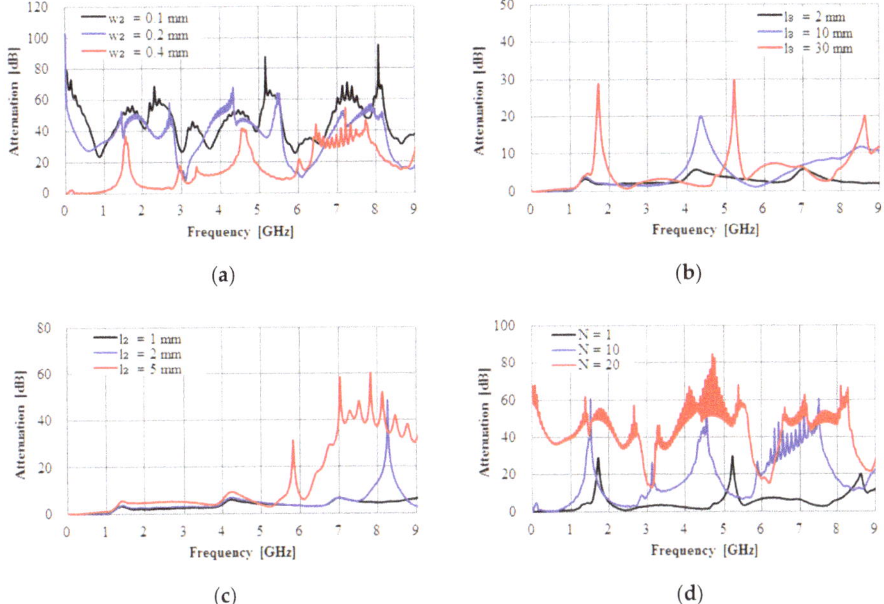

Figure 8. Lateral wave attenuation of the unit cell of the meander slot line obtained by Figure 4. (**a**) Split gap width w_2 changed (when l_1 = 0.25 mm, l_2 = 0.4 mm, l_3 = 30 mm, w_1 = 2.1 mm, $w_2 = w_3$); (**b**) length l_3 changed (when l_1 = 0.25 mm, l_2 = 0.4 mm, w_1 = 2.1 mm, $w_2 = w_3$ = 0.2 mm); (**c**) length l_2 changed (when l_1 = 0.25 mm, l_3 = 2 mm, w_1 = 2.1 mm, $w_2 = w_3$ = 0.2 mm); (**d**) number of cells changed (when l_1 = 0.25 mm, l_2 = 0.4 mm, l_3 = 30 mm, w_1 =2.1 mm, $w_2 = w_3$ = 0.2 mm.

3.3. Time Domain Crosstalk Simulation Results

The crosstalk can be explained by the current distribution on the printed circuit board. We used CST Microwave studio to obtain the current distributions of the two types of circuit boards. Figure 9 shows the surface current distribution at 2 GHz. In the case of the straight split gap, currents flow along the edge of the split and reflect at the open ends of the PCB, which enhances coupling between the two parallel signal lines and causes crosstalk. For the meander split gap, most currents cross over the meander split gap near the signal line, which shows the effectiveness of the proposed structure. Following the meander DGS split gap, the increased impedance caused by currents with alternating directions causes the crosstalk signal to be reduced effectively.

The crosstalk behavior of the transmission line equivalent circuit model for the meander DGS split is simulated in the time domain by using Agilent ADS in Figure 10. We excited a source pulse voltage $1V_{p\text{-}p}$ with 50 ohm and checked the voltage of V_3 (port #3) and V_4 (port #4). Figure 11 shows when the fast pulse signal is excited, the proposed meander DGS split decreases the crosstalk voltage at port #3 and #4 over 50% more than the straight split.

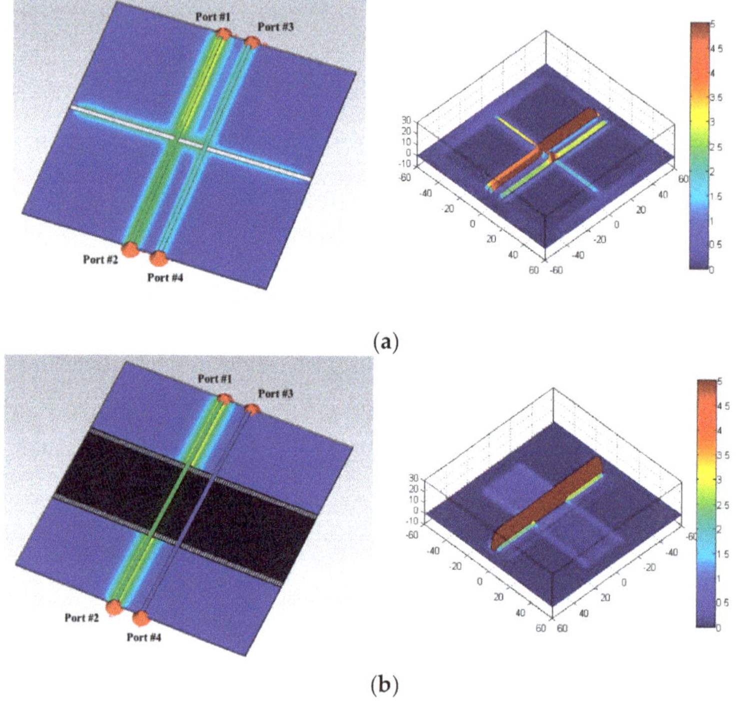

(a)

(b)

Figure 9. Current distributions on the straight and meander split power or ground plane at 2 GHz. (**a**) Straight split plane; (**b**) meander split plane.

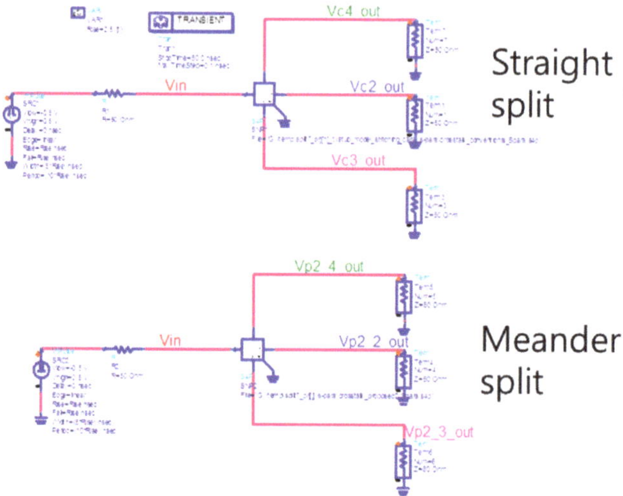

Figure 10. Time simulation of the crosstalk signal of the straight and meander split ground plane.

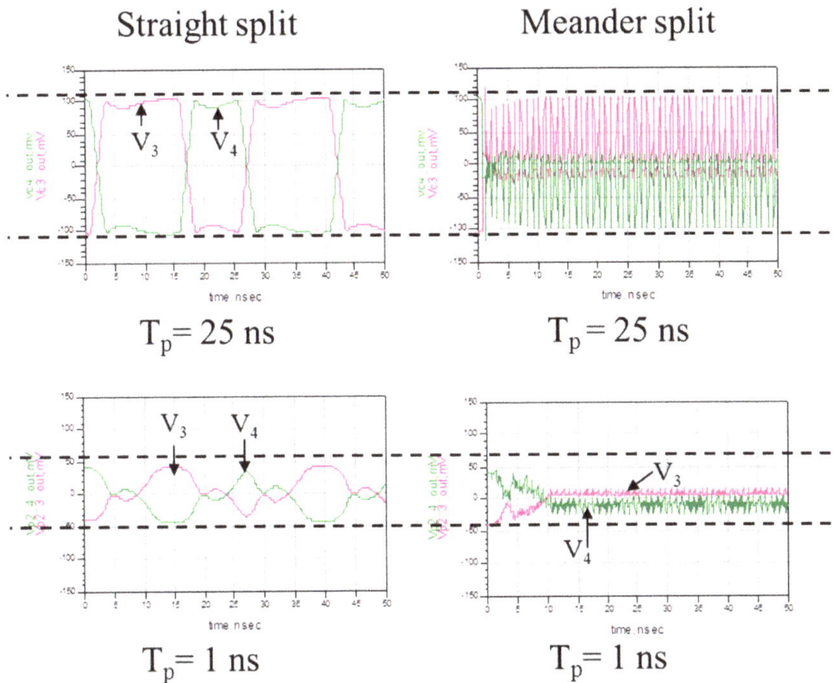

Figure 11. Time domain simulation result of Figure 10.

3.4. Time Domain Crosstalk Measurement Results

In order to demonstrate the effectiveness of the proposed split method, the crosstalk behavior of the straight and proposed split was observed experimentally in the time domain. The measurements were made by a digital oscilloscope (Tektronix TDS 6604B) and a RF signal generator (Rohde&Schwarz SML 03). The characteristic of the input RF signal is as follows. The frequency is 2 GHz, amplitude is 0.8 V. The input port #1 is fed into the RF signal generator in Figure 1. All the other ends of the lines such as port #2 (transmission), port #3 (near-end crosstalk) and port #4 (far-end crosstalk) are fed into the digital oscilloscope. Figure 12 shows the peak to peak voltage of the transmission signal of the meander split is larger than the straight split and it reduces near/far end crosstalk by over 50%. This behavior demonstrates that the proposed meander split power/ground plane can effectively eliminate the reflection and crosstalk due to a split gap and increase transmission bandwidth in the time domain.

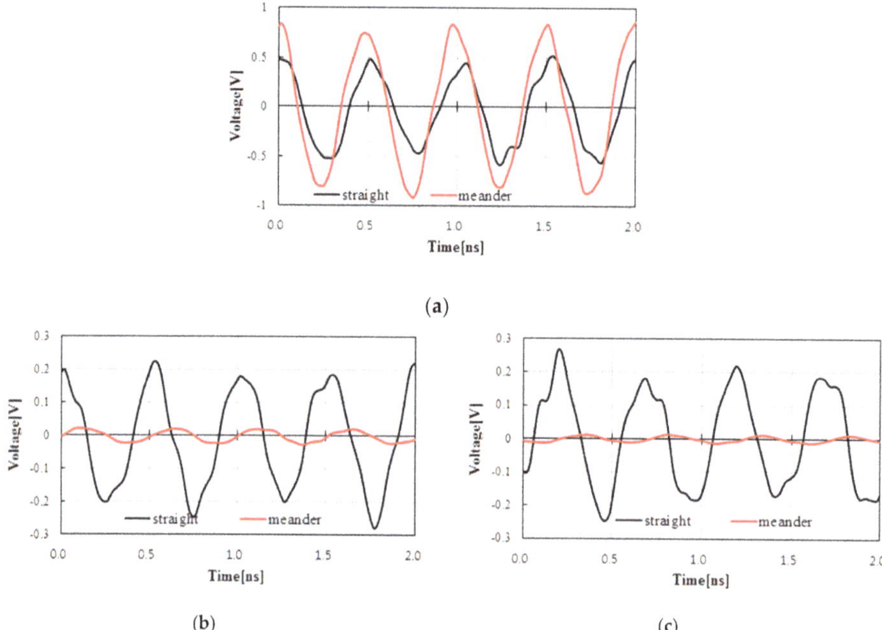

Figure 12. Comparison of the transmission and crosstalk signal with the straight and meander split power or ground plane. (**a**) Transmission signal measured at port #2; (**b**) near-end crosstalk signal measured at port #3; (**c**) far-end crosstalk signal measured at port #4.

4. Conclusions

A novel meander DGS is proposed to improve the transmission bandwidth and reduce the crosstalk of the two parallel microstrip lines over split power/ground planes. It employs a meander split to enhance electromagnetic coupling and suppress propagation of lateral waves to make signal return current flow across the split formed in the power/ground plane along the signal line in proximity, which helps reduce the crosstalk effectively. We have presented a design method for the main parameters l_3 and w_2 of the meander DGS that enhanced electromagnetic coupling and reduced the crosstalk. We developed the LC equivalent circuit and transmission line equivalent circuit model for analyzing the crosstalk. The validity of the equivalent circuit model was verified by comparison with measurement and simulation results and it shows overall good agreement with slight overestimation due to the mutual inductance interpolation error. Experimental results show that the proposed meander DGS improves signal quality and reduces near/far end crosstalk over 30 dB up to 9 GHz and 50% in the frequency domain and time domain, respectively.

Author Contributions: Conceptualization, methodology, investigation, writing-original draft preparation, writing-review and editing, J.-H.L.

Funding: This research was funded by Korea Institute of Ocean Science & Technology, grant number PE99741.

Acknowledgments: This research was part of the project titled "Development of maritime defense and security technology [PE99741]" funded by Korea Institute of Ocean Science & Technology in 2019.

Conflicts of Interest: The author declares no conflict of interest.

References

1. Fan, J.; Ren, Y.; Chen, J.; Hockanson, D.M.; Shi, H.; Drewniak, J.L.; Hubing, T.H.; Van Doren, T.P.; DuBroff, R.E. RF isolation using power islands in DC power bus design. In Proceedings of the 1999 IEEE International Symposium on Electromagnetic Compatability. Symposium Record (Cat. No. 99CH36261), Seattle, WA, USA, 2–6 August 1999; pp. 838–843.
2. Miller, J.R. The impact of split power planes on package performance. In Proceedings of the 2001 Proceedings. 51st Electronic Components and Technology Conference (Cat. No. 01CH37220), Orlando, FL, USA, 29 May–1 June 2001; pp. 1117–1121.
3. Liaw, H.-J.; Merkelo, H. Signal integrity issues at split ground and power planes. In Proceedings of the 1996 Proceedings 46th Electronic Components and Technology Conference, Orlando, FL, USA, 28–31 May 1996; pp. 752–755.
4. Shim, H.-Y.; Kim, J.; Yook, J.-G. Modeling of ESD and EMI problems in split multi-layer power distribution network. In Proceedings of the 2003 IEEE Symposium on Electromagnetic Compatibility. Symposium Record (Cat. No. 03CH37446), Boston, MA, USA, 18–22 August 2003; pp. 48–51.
5. Cui, W.; Fan, J.; Ren, Y.; Shi, H.; Drewniak, J.; Dubroff, R. DC power-bus noise isolation with power-plane segmentation. *IEEE Trans. Electromagn. Compat.* **2003**, *45*, 436–443. [CrossRef]
6. Johnson, H.W.; Graham, M. *High-Speed Digital Design: A Handbook of Black Magic*; Prentice Hall: Upper Saddle River, NJ, USA, 1993.
7. Shuppert, B. Microstrip/slotline transitions: Modeling and experimental investigation. *IEEE Trans. Microwave Theory Tech.* **1988**, *36*, 1272–1282. [CrossRef]
8. Duan, D.-W.; Rubin, B.J.; Magerlein, J. Distributed effects of a gap in power/ground planes. In Proceedings of the IEEE 8th Topical Meeting on Electrical Performance of Electronic Packaging (Cat. No. 99TH8412), San Diego, CA, USA, 25–27 October 1999; pp. 207–210.
9. Schuster, C.; Fichtner, W. Parasitic modes on printed circuit boards and their effects on EMC and signal integrity. *IEEE Trans. Electromagn. Compat.* **2001**, *43*, 416–425. [CrossRef]
10. Moran, T.; Virga, K.; Aguirre, G.; Prince, J. Methods to reduce radiation from split ground planes in RF and mixed signal packaging structures. *IEEE Trans. Adv. Packag.* **2002**, *25*, 409–416. [CrossRef]
11. Xiao, F.; Nakada, Y.; Murano, K.; Kami, Y. Crosstalk analysis model for traces crossing split ground plane and its reduction by stitching capacitor. *Electron. Commun. Jpn. (Part II: Electron.)* **2007**, *90*, 26–34. [CrossRef]
12. Suntives, A.; Khajooeizadeh, A.; Abhari, R. Using via fences for crosstalk reduction in PCB circuits. In Proceedings of the 2006 IEEE International Symposium on Electromagnetic Compatibility, Portland, OR, USA, 14–18 August 2006; pp. 34–37.
13. Mbairi, F.D.; Siebert, W.P.; Hesselbom, H. On The Problem of Using Guard Traces for High Frequency Differential Lines Crosstalk Reduction. *IEEE Trans. Compo. Packag. Technol.* **2007**, *30*, 67–74. [CrossRef]
14. Huang, W.-T.; Lu, C.-H.; Lin, D.-B. SUPPRESSION OF CROSSTALK USING SERPENTINE GUARD TRACE VIAS. *Prog. Electromagn. Res.* **2010**, *109*, 37–61. [CrossRef]
15. Koo, S.-K.; Lee, H.; Park, Y.B. Crosstalk Reduction Effect of Asymmetric Stub Loaded Lines. *J. Electromagn. Waves Appl.* **2011**, *25*, 1156–1167. [CrossRef]
16. Chu, S.T.; Little, B.; Pan, W.; Kaneko, T.; Kokubun, Y. Cascaded microring resonators for crosstalk reduction and spectrum cleanup in add-drop filters. *IEEE Photonics Technol. Lett.* **1999**, *11*, 1423–1425.
17. Cai, Q.-M.; Yu, X.-B.; Zhang, L.; Zhu, L.; Zhang, C.; Ren, Y.; Fan, J. Far-End Crosstalk Mitigation in DDR5 Using Graphene-Paraffin Material Coated Signal Lines with Tabs. In Proceedings of the 2019 Cross Strait Quad-Regional Radio Science and Wireless Technology Conference (CSQRWC 2019), Taiyuan, China, 18–21 July 2019; pp. 1–3.
18. Zhang, L.; Cai, Q.-M.; Yu, X.-B.; Zhu, L.; Zhang, C.; Ren, Y.; Fan, J. Far-End Crosstalk Mitigation for Microstrip Lines in High-Speed PCBs. In Proceedings of the 2019 Cross Strait Quad-Regional Radio Science and Wireless Technology Conference (CSQRWC 2019), Taiyuan, China, 18–21 July 2019; pp. 1–3.
19. Lee, S.-K.; Lee, K.; Park, H.-J.; Sim, J.-Y. FEXT-eliminated stub-alternated microstrip line for multi-gigabit/second parallel links. *Electron. Lett.* **2008**, *44*, 272–273. [CrossRef]
20. Roden, J.A.; Archambeault, B.; Lyle, R.D. Effect of stitching capacitor distance for critical traces crossing split reference planes. In Proceedings of the 2003 IEEE Symposium on Electromagnetic Compatibility. Symposium Record (Cat. No. 03CH37446), Boston, MA, USA, 18–22 August 2003; pp. 703–707.

21. Lin, D.-B.; Shen, R.-H. The improvement of signal quality and far-end crosstalk for coupled microstrip line over a completely split ground plane. In Proceedings of the 2012 IEEE Electrical Design of Advanced Packaging and Systems Symposium (EDAPS 2012), Taipei, Taiwan, 9–11 December 2012; pp. 139–141.
22. Kim, J.; Lee, H.; Kim, J. Effects on signal integrity and radiated emission by split reference plane on high-speed multilayer printed circuit boards. *IEEE Trans. Adv. Packag.* **2005**, *28*, 724–735.
23. Chuang, H.-H.; Wu, T.-L. Suppression of Common-Mode Radiation and Mode Conversion for Slot-Crossing GHz Differential Signals Using Novel Grounded Resonators. *IEEE Trans. Electromagn. Compat.* **2011**, *53*, 429–436. [CrossRef]
24. Chuang, H.-H.; Chou, C.-C.; Chang, Y.-J.; Wu, T.-L. A Branched Reflector Technique to Reduce Crosstalk between Slot-Crossing Signal Lines. *IEEE Microwave Wireless Compon. Lett.* **2012**, *22*, 342–344. [CrossRef]
25. Mandal, M.K.; Sanyal, S. A novel defected ground structure for planar circuits. *IEEE Microwave Wireless Compo. Lett.* **2006**, *16*, 93–95. [CrossRef]
26. Chen, H.-J.; Huang, T.-H.; Chang, C.-S.; Chen, L.-S.; Wang, N.-F.; Wang, Y.-H.; Houng, M.-P. A novel cross-shape DGS applied to design ultra-wide stopband low-pass filters. *IEEE Microwave Wireless Components Lett.* **2006**, *16*, 252–254. [CrossRef]
27. Woo, D.-J.; Lee, T.-K.; Lee, J.-W.; Pyo, C.-S.; Choi, W.-K. Novel U-slot and V-slot DGSs for bandstop filter with improved Q factor. *IEEE Trans. Microwave Theory Tech.* **2006**, *54*, 2840–2847.
28. Kim, C.-S.; Park, J.-S.; Ahn, D.; Jae-Bong, L. A novel 1-D periodic defected ground structure for planar circuits. *IEEE Microwave Guided Wave Lett.* **2000**, *10*, 131–133.
29. Balalem, A.; Ali, A.R.; Machac, J.; Omar, A. Quasi-Elliptic Microstrip Low-Pass Filters Using an Interdigital DGS Slot. *IEEE Microwave Wireless Compon. Lett.* **2007**, *17*, 586–588. [CrossRef]
30. Liu, W.-T.; Tsai, C.-H.; Han, T.-W.; Wu, T.-L. An Embedded Common-Mode Suppression Filter for GHz Differential Signals Using Periodic Defected Ground Plane. *IEEE Microwave Wireless Compon. Lett.* **2008**, *18*, 248–250.
31. Henridass, A.; Sindhadevi, M.; Karthik, N.; Alsath, M.G.N.; Kumar, R.R.; Malathi, K. Defective ground plane structure for broadband crosstalk reduction in PCBs. In Proceedings of the 2012 International Conference on Computing, Communication and Applications, Dindigul, India, 22–24 February 2012; pp. 1–5.
32. Khandelwal, M.K.; Kanaujia, B.K.; Kumar, S. Defected Ground Structure: Fundamentals, Analysis, and Applications in Modern Wireless Trends. *Int. J. Antennas Propag.* **2017**, *2017*, 1–22. [CrossRef]
33. Curnow, H. A General Equivalent Circuit for Coupled-Cavity Slow-Wave Structures. *IEEE Trans. Microwave Theory Tech.* **1965**, *13*, 671–675. [CrossRef]
34. Abbosh, A.M.; Bialkowski, M.E. Deriving characteristics of the slotline using the conformal mapping technique. In Proceedings of the MIKON 2008-17th International Conference on Microwaves, Radar and Wireless Communications, Wroclaw, Poland, 19–21 May 2008; pp. 1–4.

Article

A Dual-Perforation Electromagnetic Bandgap Structure for Parallel-Plate Noise Suppression in Thin and Low-Cost Printed Circuit Boards

Myunghoi Kim

Department of Electrical, Electronic, and Control Engineering, and the Institute for Information Technology Convergence, Hankyong National University, Anseong 17579, Korea; mhkim80@hknu.ac.kr; Tel.: +82-31-670-5295

Received: 27 May 2019; Accepted: 21 June 2019; Published: 25 June 2019

Abstract: In this study, we propose and analyze a dual-perforation (DP) technique to improve an electromagnetic bandgap (EBG) structure in thin and low-cost printed circuit boards (PCBs). The proposed DP–EBG structure includes a power plane with a square aperture and a patch with an L-shape slot that overcomes efficiently the problems resulting from the low-inductance and the characteristic impedance of the EBG structure developed for parallel-plate noise suppression in thin PCBs. The effects of the proposed dual-perforation technique on the stopband characteristics and unit cell size are analyzed using an analytical dispersion method and full-wave simulations. The closed-form expressions for the main design parameters of the proposed DP–EBG structure are extracted as a design guide. It is verified based on full-wave simulations and measurements that the DP technique is a cost-effective method that can be used to achieve a size reduction and a stopband extension of the EBG structure in thin PCBs. For the same unit cell size and low cut-off frequency, the DP–EBG structure increases the stopband bandwidth by up to 473% compared to an inductance-enhanced EBG structure. In addition, the unit cell size is substantially reduced by up to 94.2% compared to the metallo–dielectric EBG structure. The proposed DP–EBG technique achieves the wideband suppression of parallel plate noise and miniaturization of the EBG structure in thin and low-cost PCBs.

Keywords: electromagnetic bandgap (EBG); dual perforation (DP); parallel-plate noise; power delivery network (PDN); printed circuit board (PCB)

---

1. Introduction

Design complexity of high-speed digital and microwave printed circuit boards (PCBs) continues to increase as high-speed PCBs are fully populated with heterogeneous circuits and associated interconnects. High-speed PCB design is a complicated and heavily constrained problem due to the requirement to ensure reliable power delivery in the presence of multiple voltage levels and optimized signal traces within restricted routing regions. Moreover, small-form factors and cost reduction are preferred. To solve this complex problem, multilayer PCB technology, including thin and low-cost dielectrics, such as epoxy–resin fiberglass (e.g., FR-4) dielectric materials, is extensively employed in high-speed digital and microwave applications. Use of a thin dielectric provides advantages of inductance reductions when signals or planes are vertically connected, and when narrow transmission lines are used for highly dense interconnections. Additionally, a thin dielectric increases the static capacitance of PCB planes, thus achieving low-plane impedance for power delivery networks [1]. This reduces the effort of designing and optimizing decoupling capacitors for noise suppressions.

However, recent circuit operations of high-speed switching and high-bandwidth data transfers generate wideband and high-frequency noise in PCB power delivery networks that cannot be reduced or suppressed by the low-impedance characteristics of thin PCBs. In particular, parallel-plate noise is a serious problem because it significantly affects system performance. Moreover, it is induced by a parallel-plate waveguide that is frequently adopted for power delivery networks in high-speed PCBs [2–6].

One of the methods used to suppress wideband and high-frequency parallel-plate noise in high-speed PCBs is a power delivery network based on an electromagnetic bandgap (EBG) structure. To-this-date, various EBG structures have been introduced [7–22]. Their characteristics of parallel-plate noise suppression are superior. The EBG structures exhibit increased levels of noise suppression over a wideband frequency range. They are easily implemented by metal patterning of conductive layers, and can thus be simply integrated into PCBs. One promising approach introduced in previously conducted researches is the EBG structure [12–22]. This technique is based on a shunt LC resonator, whereby the capacitance and inductance are respectively induced by an embedded metal patch and a via. These EBG structures have been studied extensively and a variety of cost-effective techniques have been presented for the improvement of the stopband and the miniaturization of an EBG unit cell [12–22].

To reduce a unit cell size of an EBG structure with cost-effective PCB technology, various methods using edge-located vias and inductance-enhanced patch have been proposed. The EBG structure that uses an edge-located via [12,13] increases the inductance value of the shunt LC resonator by simply moving the via to the patch edge. In inductance-enhanced EBG structures [14–18], a resonant patch is perforated with the use of various patterns, such as spiral-shaped, stub-like, I-type patterns, so that the effective inductances in the equivalent circuit of a unit cell substantially increase.

These methods efficiently increase the inductance value of a shunt LC resonator using low-cost PCB technology. Hence, decreases of the low-cutoff frequency can result in the miniaturization of EBG structures. However, the drawback of these techniques is the significant reduction of the stopband bandwidth. The stopband bandwidth is at most 1 GHz to suppress the parallel-plate noise in the low-frequency range of 1–2 GHz.

To enhance the stopband bandwidth, an EBG structure using multiple vias is presented in [19,20]. In the multivia EBG structure, the equivalent inductance of a shunt LC resonator is reduced by the parallel connection between the vias. The multivia approach substantially increases the stopband bandwidth without changing the EBG size and without increasing manufacturing cost. However, its drawback is a low-cutoff frequency which is shifted to a high frequency, and which results in the increase of the unit cell size to suppress parallel-plate noise in the low-frequency range. Furthermore, the multivia approach is less effective in thin PCBs because the inductance effect on the stopband is not dominant for thin dielectrics.

Other methods used for stopband improvements employed defected ground structures (DGSs) [21–24]. The plane that is connected to a resonant patch through a via is etched by particular patterns so that the characteristic impedance (Z_o) increases in an equivalent EBG unit cell circuit. A stopband bandwidth significantly increases, while a low-cutoff frequency of the DGS–EBG structure is not shifted to a high frequency. While its stopband expansion is prominent, the DGS–EBG structure does not have the advantage of miniaturizing an EBG structure in thin PCBs. Consequently, it is necessary to develop a new technique to simultaneously achieve increases of the stopband bandwidth and size reductions of the EBG structure in thin PCBs.

In this study, a dual-perforation technique is proposed for a miniaturized and wideband EBG structure to mitigate parallel-plate noise in thin and low-cost PCBs. The study is organized as follows: (1) in Section 2, the proposed EBG structure is presented, and its improved features are completely explained using dispersion analysis based on the close-form expressions for low and high cut-off frequencies and full-wave simulation approaches based on finite element method. (2) In Section 3, the proposed EBG structure is validated using the scattering parameters which are obtained using the full-wave simulations and experimental results. (3) The conclusions of the study are outlined in Section 4.

2. Dual-Perforation EBG Structure

2.1. Design Description

The proposed dual-perforation EBG (DP–EBG) structure is developed to suppress parallel-plate noise in multilayer PCBs, including thin dielectrics. The DP–EBG structure is a periodic structure in which a unit cell comprises a perforated power plane, a perforated patch, and a ground plane embedded in a thin dielectric, as shown in Figure 1. The power plane is perforated by four square apertures, while the resonant patch is perforated by an L-shape slot. These perforations are simple to implement with conventional PCB manufacturing techniques without requiring additional, costly processes. The perforated power plane and patch are connected through a short-length via owing to the thin dielectric. The apertures in the power plane electrically enhance the characteristic impedance of the EBG unit cell, while the L-shape slot in the resonant patch effectively increases the inductance of the EBG unit cell. The unit cell size and square aperture of the DP–EBG structure are represented as d_c–by–d_c and d_a–by–d_a structures, respectively. Therefore, the width of the remaining conductor is the same and equal to d_c-2d_a. Regarding the L-shape slot, the conductor width etched on the patch is denoted by w_s and the width of the remaining conductor is denoted by w_b. The sum of $d_p/2$ and d_s is the total length of the L-shape slot. The design parameter d_s is adjustable to obtain a desired patch inductance. The dielectric thickness between the dual perforated planes is h_1. The distance between the perforated patch and the ground plane is h_2.

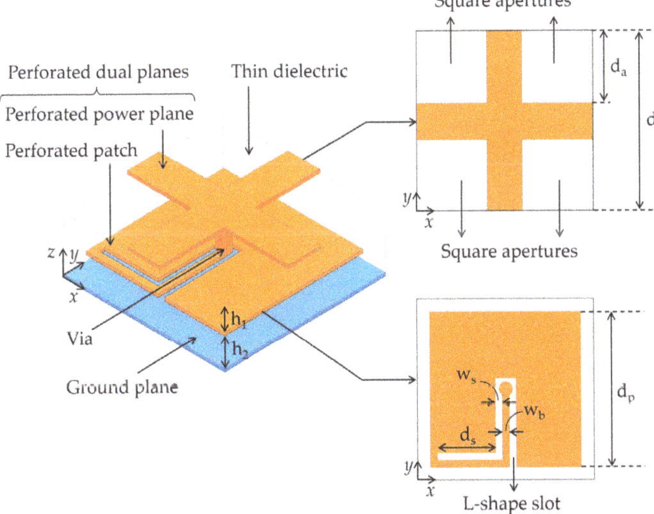

Figure 1. A unit cell of a dual-perforation electromagnetic bandgap dual-perforation–electromagnetic bandgap (DP–EBG) structure consisting of perforated dual planes and its design parameters.

2.2. Derivation of the Equations for f_L and f_H

Dispersion analysis is performed to examine the dual-perforation effects on the stopband of the DP–EBG structure. To derive an analytical dispersion equation, a two-port equivalent circuit model based on transmission line theory is considered, as shown in Figure 2. In this study, one-dimensional (1–D) propagation is captured by the equivalent circuit. The dispersion analysis conducted with the use of the 1–D circuit model can be extended to a 2–D EBG array because it shows good correlation with the prediction of noise suppression in the 2–D EBG array. The DP–EBG circuit model consists of two transmission lines and a resonator circuit in which an inductor and a capacitor are connected in series. At the resonant frequency, the LC resonator has zero impedance which results in the decoupling of parallel-plate noise. The transmission line is the circuital representation of electromagnetic waves propagating through the parallel plate waveguide, which is formed by the perforated power and the ground planes. In the DP–EBG circuit model, Z_{eq} and β_c respectively denote the characteristic impedance and propagation constant of this parallel plate waveguide or transmission line. Its length is $d_c/2$. In the transmission line model, Z_{eq} is significantly enhanced by the square perforation aperture because the width of the remaining conductor perpendicular to the direction of wave propagation is narrowed. In the resonator circuit, the capacitance C_p is attributed to the capacitor between the resonant patch and the corresponding ground plane, while the inductance is attributed to the patch and the via inductances. For the DP–EBG structure in the thin PCBs, the inductance of the via can be ignored compared to the patch inductance which substantially increases due to the L-shape slot. Thus, the value of L_{eq} in the DP–EBG circuit model is mainly determined by the L-shape slot on the perforated resonant patch. Hence, the equivalent circuit of the proposed DP–EBG structure is expressed with a Z_o-enhanced transmission line and an L-enhanced resonator.

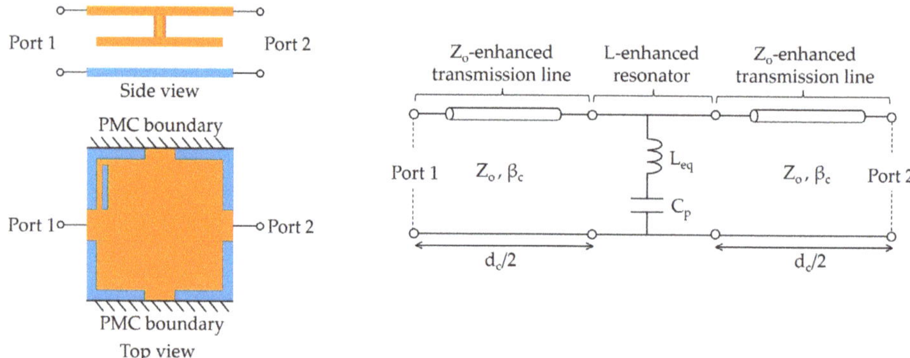

Figure 2. Equivalent circuit model of the DP–EBG unit cell based on transmission line theory for the extraction of the dispersion equation.

Using the ABCD parameters of the microwave theory, the voltage/current relationships between ports 1 and 2 of the DP–EBG unit cell are described by [22]

$$\begin{pmatrix} \cos(\beta_{uc}d_c) & jZ_{uc}\sin(\beta_{uc}d_c) \\ jZ_{uc}^{-1}\sin(\beta_{uc}d_c) & \cos(\beta_{uc}d_c) \end{pmatrix} = \begin{pmatrix} \cos\left(\frac{\beta_c d_c}{2}\right) & jZ_{eq}\sin\left(\frac{\beta_c d_c}{2}\right) \\ jZ_{eq}^{-1}\sin\left(\frac{\beta_c d_c}{2}\right) & \cos\left(\frac{\beta_c d_c}{2}\right) \end{pmatrix} \begin{pmatrix} 1 & 0 \\ Y_R & 1 \end{pmatrix} \begin{pmatrix} \cos\left(\frac{\beta_c d_c}{2}\right) & jZ_{eq}\sin\left(\frac{\beta_c d_c}{2}\right) \\ jZ_{eq}^{-1}\sin\left(\frac{\beta_c d_c}{2}\right) & \cos\left(\frac{\beta_c d_c}{2}\right) \end{pmatrix} \quad (1)$$

where,

$$Y_R = \frac{j(2\pi f)C_p}{1-(2\pi f)^2 L_{eq}C_p} \quad (2)$$

From the equations listed above, an effective phase constant β_{uc} of the DP–EBG unit cell is derived by

$$\beta_{uc} = \frac{1}{d_c}\cos^{-1}\left[\cos(\beta_c d_c) - \frac{(2\pi f)C_p Z_{eq}}{2\left(1-(2\pi f)^2 C_p L_{eq}\right)}\sin(\beta_c d_c)\right]. \quad (3)$$

Closed-form expressions for low and high-cutoff frequencies (f_L and f_H) are further extracted from (3). To derive a closed-form expression for f_L, it is considered that the real part of β_{uc} is equal to π/d_c (i.e., $Re\{\beta_{uc}\} = \pi/d_c$). It is assumed that the electrical length of $\beta_c d_c$ for f_L is small enough to set cos($\beta_c d_c$) and sin($\beta_c d_c$) to 1 and $\beta_c d_c$, respectively. Accordingly, the following equation is obtained from (3),

$$1 = \left(\frac{\pi f_L C_p Z_{eq}}{2\left(1-(2\pi f_L)^2 C_p L_{eq}\right)}\right)\left(\frac{2\pi f_L d_c}{v_p}\right). \quad (4)$$

where v_p is the phase velocity of $c/\sqrt{\varepsilon_r}$, and c is the speed of light in vacuum. Based on Equation (4), an analytical equation can be derived explicitly for f_L, as follows,

$$f_L = \frac{1}{2\pi}\left(\frac{1}{4Z_{eq}C_p d_c v_p^{-1} + L_{eq}C_p}\right)^{1/2} \quad (5)$$

As it can be observed in Equation (5), f_L is expected to be reduced when Z_{eq} and L_{eq} increase, when the DP technique is used, which are associated with the perforations of a power plane and a resonant patch. L_{eq} mainly contributes to the reduction in f_L, while the Z_{eq} effect is limited because it is divided by the phase velocity v_p.

To obtain an explicit expression for f_H, it is considered that $\beta_{uc}d_c = 0$ or cos($\beta_{uc}d_c$) = 1. Thus, Equation (3) may be simplified to

$$tan\left(\frac{\beta_c d_c}{2}\right) = -\frac{(2\pi f_H)C_p Z_{eq}}{2\left(1-(2\pi f_H)^2 C_p L_{eq}\right)} \quad (6)$$

Accordingly, Equation (6) can be approximated using two assumptions. First, d_c is so small compared to the wavelength of f_H which results in tan($\beta_c d_c/2$) ≈ $\beta_c d_c/2$.

Second, f_H is higher than the resonant frequency determined by the $C_p L_{eq}$ product. Thus, (1-$(2\pi f_H)^2 C_p L_{eq}$) is approximated to be equal to -$(2\pi f_H)^2 C_p L_{eq}$. Consequently, (6) becomes

$$f_H = \frac{1}{2\pi}\left(\frac{Z_{eq}}{L_{eq}}\frac{v_p}{d_c}\right)^{1/2} \quad (7)$$

It is observed in Equation (7) that f_H can be shifted to lower the frequency values by increasing L_{eq} with the resonant patch perforated by the L-shape slot. Considering the factor L_{eq} in Equations (6) and (7), the f_H reduction rate is higher than that of f_L because f_H is inversely proportional to $(L_{eq})^{1/2}$. However, increasing Z_{eq}, induced from the power plane perforated using rectangular apertures, compensates this bandwidth reduction in the proposed DP–EBG structure.

2.3. Dispersion Analysis

To validate the f_L and f_H equations and examine the DP technique effects on the stopband characteristics of the DP–EBG structure, a full-wave simulation based on a finite element method (FEM) is adopted, and the results are compared with those from Equations (5) and (7). FEM simulations were performed using the commercial software HFSS (ver. 17.1, Ansys. Corp., Pittsburgh, PA, USA). The FEM simulation model used for the dispersion analysis of the DP–EBG structure is identical to the unit cell shown in Figure 1. In the simulation model, the nominal values of the design parameters of d_c, d_p, d_a, w_s, w_b, d_s, h_1, and h_2, are set to 5 mm, 4.9 mm, 2.45 mm, 0.1 mm, 0.1 mm, 2.2 mm, 0.1 mm, and 0.1 mm, respectively. These values were determined based on the consideration of a conventional and low-cost PCB process. The dimensions of the geometrical parameters are summarized in Table 1. The dielectric thicknesses of h_1 and h_2 are chosen as the minimum value provided by the cost-effective PCB process. The dielectric constant and loss tangent of the FR–4 are 4.4 and 0.02, respectively. The aperture size d_a and the L-shape slot length d_s are respectively related with Z_{eq} and L_{eq}. The parameters d_a and d_s were varied to comprehensively examine the stopband characteristics of the DP–EBG structure and to verify the closed-form expressions for f_L and f_H of Equations (5) and (7). The d_a values of 2.15, 2.25, 2.35, and 2.45 mm, are employed which correspond to the Z_{eq} values of 33,

41, 56, and 90 Ω. The d_s values changed from 0.15 mm to 2.3 mm and coincided with the L_{eq} values from 1.55 nH to 2.38 nH. The Z_{eq} and L_{eq} values associated with the geometrical dimensions were simply obtained using quasistatic simulations.

Table 1. Dimensions of geometrical parameters of the DP–EBG structure.

Parameters	d_c	d_p	d_a	w_s	w_b	d_s	h_1	h_2
Dimensions (mm)	5	4.9	2.45	0.1	0.1	2.2	0.1	0.1

Figure 3 depicts the f_L and f_H values of the stopband characteristics with the aforementioned Z_{eq} and L_{eq} values, as acquired from the FEM simulations (blue lines) and the proposed equations (red lines). The results for Z_{eq} of 33, 41, 56, and 90 Ω, are shown in Figure 3a–d, respectively. The closed-form expressions for f_L and f_H exhibit good correlations with the FEM-based full-wave simulations, as shown in all the figures. The discrepancies associated with f_H may result from the first-order approximation of the Taylor series expansion of the tangent function. However, the differences between the FEM simulation and the closed-form expressions are approximately uniform for all the L_{eq} values. Thus, the tendencies among these results are in close agreement. Even though the closed-form expressions for f_L and f_H derived herein are verified using a limited number of test cases, these equations can be extended and applied to other DP–EBG structures, including the different dimensions of geometrical parameters.

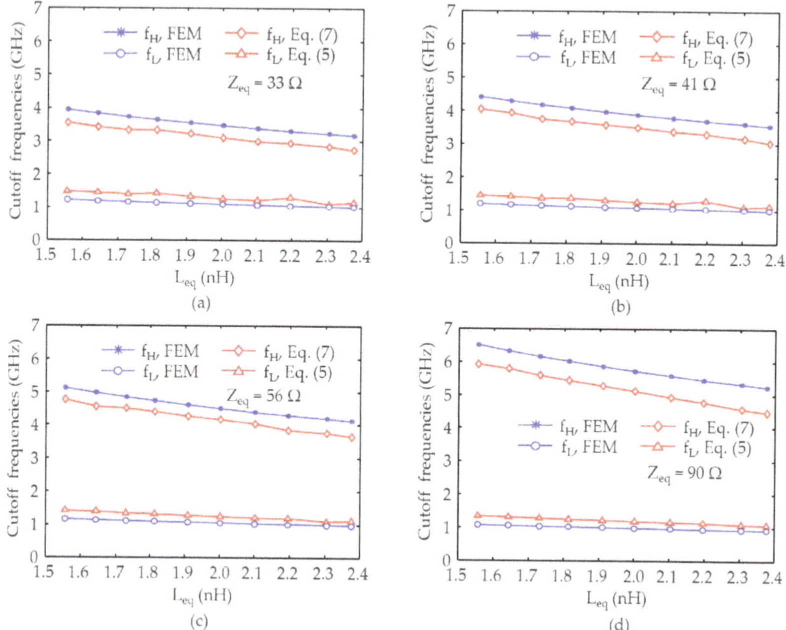

Figure 3. Various cutoff frequencies with respect to the changes of Z_{eq} ((**a**) 33 Ω, (**b**) 41 Ω, (**c**) 56 Ω and (**d**) 90 Ω) and L_{eq} used to examine the stopband characteristics and verify the closed-form expressions for f_L and f_H.

The effects of the DP technique on a stopband are further examined. The f_L and f_H variations with respect to Z_{eq} and L_{eq} are explored using Equations (5) and (7), as shown in Figure 4. The value of Z_{eq} varies from 5 Ω to 95 Ω and that of L_{eq} changes from 1.0 nH to 3.0 nH. These values are commonly used for the proposed DP–EBG structure in low-cost and thin PCBs. The overall tendencies of the variations of f_L and f_H associated with the DP technique are also observed and evaluated. In Figure 4a, the minimum f_L value is 0.85 GHz when Z_{eq} and L_{eq} are 95 Ω and 3.0 nH, respectively. The maximum value of f_L is 1.63 GHz and results from Z_{eq} = 5 Ω and L_{eq} = 1.0 nH. As shown in Figure 4b, the minimum value of f_H is 1.10 GHz and is observed when Z_{eq} = 5 Ω and L_{eq} = 3.0 nH, while the DP–EBG structure has a maximum value of f_H is 8.34 GHz when Z_{eq} = 95 Ω and L_{eq} = 1.0 nH. The conditions for the minimum and maximum f_L and f_H values are different.

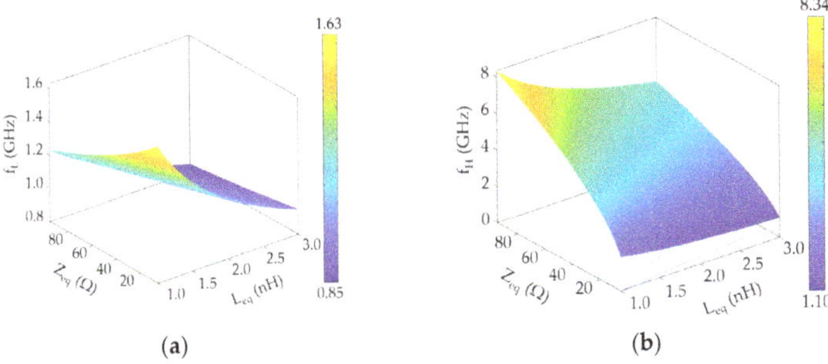

Figure 4. Variations of (**a**) f_L and (**b**) f_H for various Z_{eq} and L_{eq} values obtained using the closed-form expressions.

To gain insight into the efficient design of the DP–EBG structure, the variations of f_L and f_H are presented graphically. Figure 5 depicts the contour lines of both f_L (black dashed lines) and f_H (blue solid lines) when Z_{eq} changes from 5 Ω to 95 Ω and when L_{eq} changes from 1.0 nH to 3.0 nH. Point A in Figure 5 shows that f_L and f_H are equal to 1.23 GHz and 4.42 GHz when Z_{eq} and L_{eq} are 40 Ω and 1.5 nH, respectively. However, the suppression region of the parallel-plate noise given at point A needs to be extended in the low-frequency range. To achieve this, L_{eq} can increase. For instance, f_L changes from 1.23 GHz to 1.0 GHz as L_{eq} increases from 1.5 nH (point A) to 2.42 nH (point B) by maintaining Z_{eq} to 40 Ω. In this approach, both f_L and f_H are lowered, thus reducing the stopband bandwidth. To compensate for this drawback, other approaches can be considered, namely, by moving point A to C. Z_{eq} increases from 40 Ω to 60 Ω and L_{eq} increases from 1.5 nH to 2.42 nH, thus resulting in an f_L value of 1.0 GHz and an f_H value of 4.42 GHz. The f_L reduction is successfully achieved by maintaining f_H to 4.42 GHz. Consequently, the suppression region of the parallel plate noise is broadened as is f_L is reduced. As it has been observed in the proposed analysis, the contour plots, which were extracted with the use of the closed-form expressions for f_L and f_H, provided a simple and systematic approach to design the DP–EBG structure in thin and cost-effective PCBs.

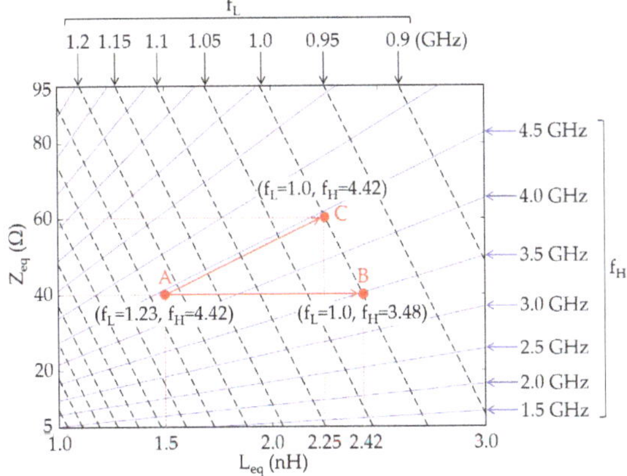

Figure 5. Contour plots of f_L and f_H used for the design and analyses of DP–EBG structures.

2.4. Performance Comparisons

The performances of the proposed DP–EBG structure are demonstrated by comparing their dispersion characteristics with those of previous EBG structures, namely a mushroom-type EBG (MT–EBG), defected-ground EBG (DGS–EBG), and inductance-enhanced EBG (IEP–EBG) structures. The unit cells and the dimensions of these EBG structures are shown in Figure 6 and Table 1, respectively. It is noted that the unit cells of these EBG structures have the same size. The dispersion characteristics are obtained by applying the Floquet theory to full-wave simulation results [25]. The results are illustrated in Figure 7. The f_L values of the previously proposed MT–EBG, DGS–EBG, IEP–EBG, and the proposed DP–EBG structures are 4.95, 2.28, 1.22, and 1.00 GHz, respectively. For the same unit cell size, the proposed DP–EBG structure shows the lowest f_L in the EBG structures. Moreover, the proposed DP–EBG structure substantially reduces f_L compared to the MT– and DGS–EBG structures. Even though the stopband bandwidth of the MT– and DGS–EBG structures are larger than the DP–EBG structure, the stopbands of the MT– and DGS–EBG structures are in higher frequency ranges. These ranges cannot be lowered unless their unit cell sizes are significantly enlarged. The IEP–EBG structure has a low f_L value comparable to the DP–EBG structure. However, the stopband bandwidth of the IEP–EBG structure is 0.48 GHz, which is equal to at most 0.17 times the stopband bandwidth of the DP–EBG structure. The DP–EBG structure successfully overcomes the limitation of the IEP–EBG structure, thus widening the bandwidth of the stopband.

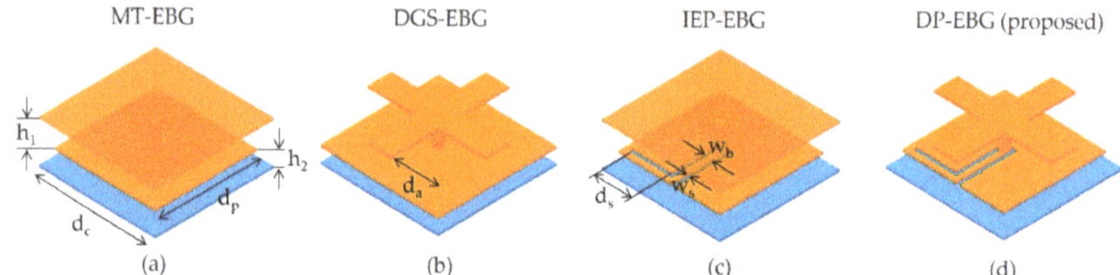

Figure 6. Unit cells of (**a**) mushroom-type (MT)–, (**b**) defected ground structure (DGS)–, (**c**) inductance-enhanced patch (IEP)–, and (**d**) DP–EBG structures for stopband comparisons.

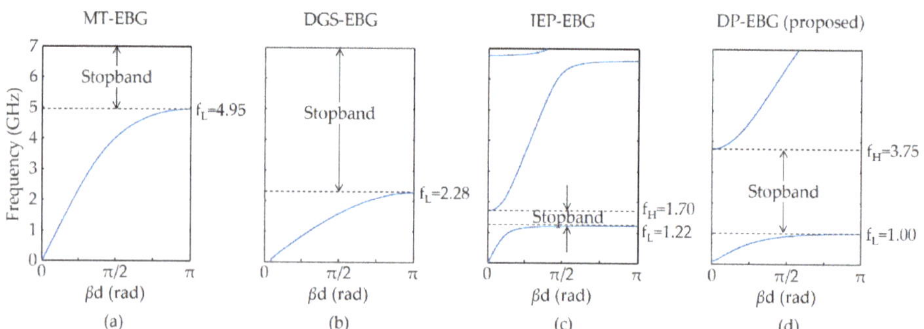

Figure 7. Dispersion diagrams of (**a**) MT–, (**b**) DGS–, (**c**) IEP–, and (**d**) DP–EBG structures.

To examine the advantage of miniaturization, an FEM-based dispersion analysis is performed to compare the previous MT–EBG and the proposed DP–EBG structures. The unit cell areas are found when the MT–EBG and DP–EBG structures contain the same f_L. The comparison result is depicted in Figure 8. An amount of unit cell area reduction substantially increases as the f_L is lowered. For an f_L value equal to 1.0 GHz, the area reduction of the DP–EBG structure is 94.2% compared to the MT–EBG structure. It is shown that the DP–EBG structure is advantageous because it downsizes the unit cell. Remarkably, this enhancement is achieved in dual-plane perforation cases only, without requiring costly materials and additional PCB processes.

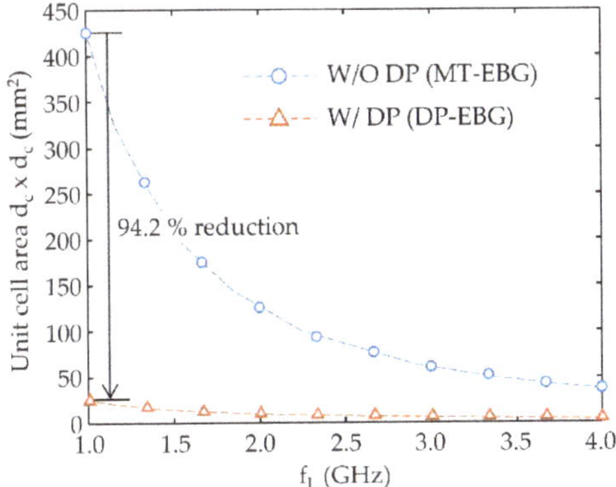

Figure 8. Comparison of unit-cell miniaturization between previous MT- and proposed DP–EBG structures.

3. Results

In this section, the parallel-plate noise suppression of the DP–EBG structure in thin PCBs is demonstrated based on the scattering parameters (S-parameters) which are obtained from the full-wave simulation of the DP–EBG structure with a 7 × 7 array. Moreover, it is experimentally verified that the DP–EBG structure suppresses parallel-plate noise in thin PCBs using a test vehicle fabricated with conventional PCB process.

3.1. Simulated Results

A full-wave simulation model of the DP–EBG structure with a 7 × 7 array is depicted in Figure 9. The array size is determined to implement the quasiperiodic condition of the DP–EBG structure. Two-port simulation is performed with waveguide ports renormalized to 50 Ω. The boundaries are set to perfect magnetic conductors and a perfect matched layer, as shown in Figure 9. To compare the noise suppression performance, the simulated S-parameters of the previous EBG structures with the 7 × 7 array and the parallel plate waveguide (PPW) without any EBG structure are also obtained. The dimensions of the geometrical parameters were described in the previous section. The port locations and the boundary conditions are identical for all EBG structures and the PPW.

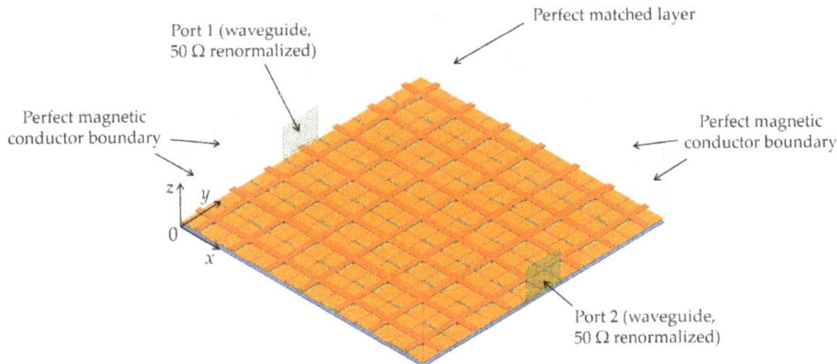

Figure 9. Finite difference method (FEM)-based simulation model of the DP–EBG structure with a 7 × 7 array.

To prove the existence of a wideband stopband with a low f_L, the simulated S_{21} parameters of the PPW, IEP–EBG, and DP–EBG structures, are shown in Figure 10a. As it can be observed, the PPW without any EBG structure is vulnerable to parallel plate noise. The stopband with a −40 dB suppression level of the IEP–EBG structure forms in the frequency range from 1.01 GHz to 1.36 GHz, while that of the DP–EBG structure ranges from 0.98 GHz to 3.32 GHz. For the same size of the EBG structures, the stopband bandwidth of the DP–EBG structure is approximately 6.7 times wider than that of the IEP–EBG structure. Moreover, the stopband of the proposed DP–EBG structure is significantly lowered compared to those of the MT–EBG and DGS–EBG structures, as shown in Figure 10b. The f_L values of the MT–EBG, DGS–EBG, and DP–EBG structures are 4.6, 3.33, and 0.98 GHz, respectively. The DP–EBG structure substantially reduces the f_L value up to 78.7% without adding costly materials and processes. The stopband estimation based on the S-parameter exhibits a good correlation with dispersion analysis results. The f_L and f_H predicted from the dispersion analysis are 1.0 and 3.75 GHz, respectively.

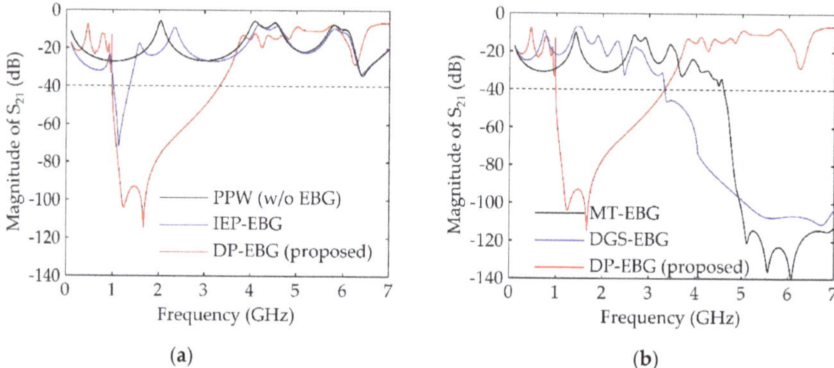

Figure 10. Comparison of S-parameters between (**a**) the ppw, IEP-EBG and proposed EBG structures and (**b**) MT-EBG, DGS-EBG, and proposed EBG structures to demonstrate a broad stopband bandwidth and miniaturization.

3.2. Measurements

To experimentally verify the DP–EBG structure, a test vehicle is fabricated using conventional PCB process. The process provides a copper-based conduction layer, FR–4 dielectric, through-hole via, and a minimum dielectric thickness of 100 μm. The via diameter is 0.4 mm and the copper thickness is 17 μm. The dielectric constant and loss tangent of the FR–4 are 4.4 and 0.02, respectively. The dimensions for the test vehicle of the DP–EBG structure are listed in Table 1. The test vehicle includes a 7 × 7 array and the entire board size is 35 mm x 35 mm. The measurement setup and fabricated PCBs of the DP–EBG structure are depicted in Figure 11. To obtain the S-parameters of the DP–EBG structure, a vector network analyzer (Anritsu MS46122A, 1 MHZ to 1 GHz) and microprobes (GSG type, 500 μm pitch) are employed. The probing pads on the test vehicle are located at (0 mm, 17.5 mm) and (35 mm, 17.5 mm) with the origin placed at the lower left corner of the PCBs. This setup is the same as the setup of the full-wave simulation described in the previous section. The measured and simulated S_{21} parameters are shown in Figure 12. The stopband of the DP–EBG structure is clearly observed in the measurement. The f_L and f_H values with a −40 dB suppression level are equal to 1.03 and 3.32 GHz, respectively. The measurements show good agreement with the full-wave simulation result. Consequently, it is experimentally verified that the DP–EBG structure substantially suppresses the parallel-plate noise with the advantage of miniaturization (in other words, low f_L) in thin and low-cost PCBs.

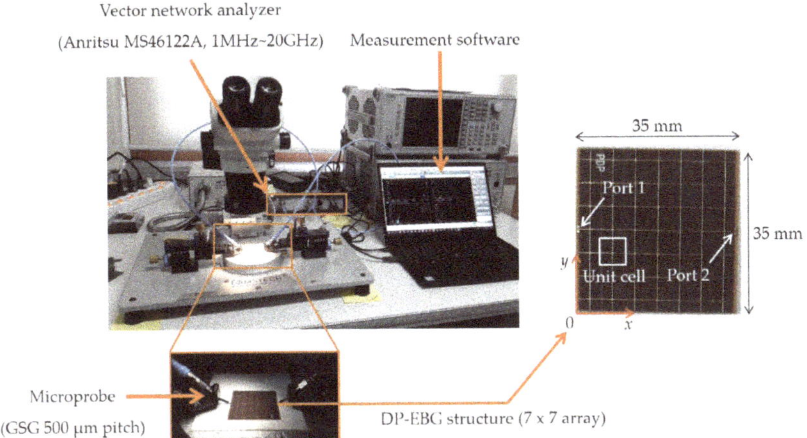

Figure 11. Measurement setup for DP–EBG structure with a 7 × 7 array.

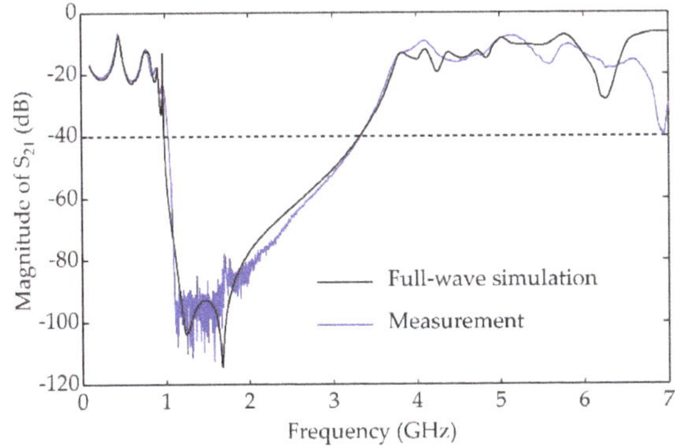

Figure 12. Measured and simulated S_{21} parameters of the DP–EBG structure.

4. Conclusions

The DP–EBG structure was proposed to improve parallel-plate noise suppression and downsize the EBG structure in multilayer PCBs with thin dielectrics. The proposed DP technique efficiently overcame the limitations of the previous EBG structure in thin and low-cost PCBs. The perforation pattern for a resonant patch lowered the start frequency of the stopband and the plane perforation improved the stopband bandwidth. The DP technique successfully achieved these without any costly materials and processes. The improved characteristics of the DP–EBG structures were thoroughly examined and validated using dispersion analysis, full-wave simulations, and experiments. In this study, the particular patterns of the rectangular aperture and L-shape slot for the DP technique are presented. Additional research studies on other patterns for the DP technique need to be conducted. It is thus anticipated that a synthesis algorithm will be developed for the various patterns planned to be tested using the DP technique.

Author Contributions: The author conceived and designed the experiments, analyzed the characteristics, performed the simulations and experiments, and wrote the paper.

Funding: This work was supported by the National Research Foundation of Korea (NRF) grant funded by the Korea government (Ministry of Science and ICT) (NRF-2019R1C1C1005777).

Conflicts of Interest: The author declares no conflict of interest.

Abbreviations

DGS	defected ground structure
DP	dual perforation
EBG	electromagnetic bandgap
FEM	finite difference method
IEP	inductance-enhanced patch
MT	mushroom-type
PCB	printed circuit board
PPW	parallel plate waveguide

References

1. Novak, I. Lossy power distribution networks with thin dielectric layers and/or thin conductive layers. *IEEE Trans. Adv. Packag.* **2000**, *23*, 353–360. [CrossRef]
2. Senthinathan, R.; Prince, J.L. Simultaneous switching ground noise calculation for packaged CMOS devices. *IEEE J. Solid-State Circuits* **1991**, *26*, 1724–1728. [CrossRef]
3. McCredie, B.D.; Becker, W.D. Modeling, measurement, and simulation of simultaneous switching noise. *IEEE Trans. Compon. Packag. Manuf. Technol. Part B* **1996**, *19*, 461–472. [CrossRef]
4. Chun, S.; Swaminathan, M.; Smith, L.D.; Srinivasan, J.; Jin, Z.; Iyer, M.K. Modeling of simultaneous switching noise in high speed systems. *IEEE Trans. Adv. Packag.* **2001**, *24*, 132–142. [CrossRef]
5. Tang, K.T.; Friedman, E.G. Simultaneous switching noise in on-chip CMOS power distribution networks. *IEEE Trans. Very Larg. Scale Integr. Syst.* **2002**, *10*, 487–493. [CrossRef]
6. Swaminathan, M.; Chung, D.; Grivet-Talocia, S.; Bharath, K.; Laddha, V.; Xie, J. Designing and Modeling for Power Integrity. *IEEE Trans. Electromagn. Compat.* **2010**, *52*, 288–310. [CrossRef]
7. Kamgaing, T.; Ramahi, O.M. A novel power plane with integrated simultaneous switching noise mitigation capability using high impedance surface. *IEEE Microw. Wirel. Compon. Lett.* **2003**, *13*, 21–23. [CrossRef]
8. Abhari, R.; Eleftheriades, G.V. Metallo-dielectric electromagnetic bandgap structures for suppression and isolation of the parallel-plate noise in high-speed circuits. *IEEE Trans. Microw. Theory Tech.* **2003**, *51*, 1629–1639. [CrossRef]
9. Lee, J.; Kim, H.; Kim, J. High dielectric constant thin film EBG power/ground network for broad-band suppression of SSN and radiated emissions. *IEEE Microw. Wirel. Compon. Lett.* **2005**, *15*, 505–507.
10. Park, J.; Lu, A.C.W.; Chua, K.M.; Wai, L.L.; Lee, J.; Kim, J. Double-stacked EBG structure for wideband suppression of simultaneous switching noise in LTCC-based SiP applications. *IEEE Microw. Wirel. Compon. Lett.* **2006**, *16*, 481–483. [CrossRef]
11. Wang, C.D.; Yu, Y.M.; de Paulis, F.; Scogna, A.C.; Orlandi, A.; Chiou, Y.P.; Wu, T.L. Bandwidth Enhancement Based on Optimized Via Location for Multiple Vias EBG Power/Ground Planes. *IEEE Trans. Compon. Packag. Manuf. Technol.* **2012**, *2*, 332–341. [CrossRef]
12. Rajo-Iglesias, E.; Inclan-Sanchez, L.; Vazquez-Roy, J.; Garcia-Munoz, E. Size Reduction of Mushroom-Type EBG Surfaces by Using Edge-Located Vias. *IEEE Microw. Wirel. Compon. Lett.* **2007**, *17*, 670–672. [CrossRef]
13. Awasthi, S.; Biswas, A. Compact filter using coupled metamaterial mushroom resonators with corner vias. In Proceedings of the 2011 IEEE Applied Electromagnetics Conference (AEMC), Kolkata, India, 18–20 December 2011; pp. 1–4.
14. Kamgaing, T.; Ramahi, O.M. Multiband Electromagnetic-Bandgap Structures for Applications in Small Form-Factor Multichip Module Packages. *IEEE Trans. Microw. Theory Tech.* **2008**, *56*, 2293–2300. [CrossRef]
15. Kasahara, Y.; Toyao, H. Compact (lambda/45-Sized) Electromagnetic Bandgap Structures With Stacked Open-Circuit Lines. *IEEE Microw. Wirel. Compon. Lett.* **2017**, *27*, 694–696. [CrossRef]
16. Kasahara, Y.; Toyao, H.; Hankui, E. Compact and Multiband Electromagnetic Bandgap Structures with Adjustable Bandgaps Derived From Branched Open-Circuit Lines. *IEEE Trans. Microw. Theory Tech.* **2017**, *65*, 2330–2340. [CrossRef]
17. Shen, C.; Chen, S.; Wu, T. Compact Cascaded-Spiral-Patch EBG Structure for Broadband SSN Mitigation in WLAN Applications. *IEEE Trans. Microw. Theory Tech.* **2016**, *64*, 2740–2748. [CrossRef]
18. Kim, M. A Miniaturized Electromagnetic Bandgap Structure Using an Inductance-Enhanced Patch for Suppression of Parallel Plate Modes in Packages and PCBs. *Electronics* **2018**, *7*, 76. [CrossRef]

19. Zhang, M.; Li, Y.; Jia, C.; Li, L. A Power Plane with Wideband SSN Suppression Using a Multi-Via Electromagnetic Bandgap Structure. *IEEE Microw. Wirel. Compon. Lett.* **2007**, *17*, 307–309. [CrossRef]
20. Park, H.H. Reduction of Electromagnetic Noise Coupling to Antennas in Metal-Framed Smartphones Using Ferrite Sheets and Multi-Via EBG Structures. *IEEE Trans. Electromagn. Compat.* **2018**, *60*, 394–401. [CrossRef]
21. Han, Y.; Huynh, H.A.; Kim, S. Pinwheel Meander-Perforated Plane Structure for Mitigating Power/Ground Noise in System-in-Package. *IEEE Trans. Compon. Packag. Manuf. Technol.* **2018**, *8*, 562–569. [CrossRef]
22. Kim, M.; Koo, K.; Hwang, C.; Shim, Y.; Kim, J.; Kim, J. A Compact and Wideband Electromagnetic Bandgap Structure Using a Defected Ground Structure for Power/Ground Noise Suppression in Multilayer Packages and PCBs. *IEEE Trans. Electromagn. Compat.* **2012**, *54*, 689–695.
23. Kim, M.; Kam, D.G. A Wideband and Compact EBG Structure with a Circular Defected Ground Structure. *IEEE Trans. Compon. Packag. Manuf. Technol.* **2014**, *4*, 496–503. [CrossRef]
24. Kim, Y.; Cho, J.; Cho, K.; Park, J.; Kim, S.; Kim, D.H.; Park, G.; Sitaraman, S.; Raj, P.M.; Tummala, R.R.; et al. Glass-Interposer Electromagnetic Bandgap Structure with Defected Ground Plane for Broadband Suppression of Power/Ground Noise Coupling. *IEEE Trans. Compon. Packag. Manuf. Technol.* **2017**, *7*, 1493–1505. [CrossRef]
25. Collin, R.E. *Field Theory of Guided Waves*; IEEE Press: New York, NY, USA, 1990.

Article

Shielding Properties of Cement Composites Filled with Commercial Biochar

Muhammad Yasir [1], Davide di Summa [2], Giuseppe Ruscica [2], Isabella Natali Sora [2] and Patrizia Savi [1,*]

1 Department of Electronics and Telecommunications, Polytechnic of Turin, C.so Duca degli Abruzzi 24, 10129 Torino, Italy; muhammad.yasir@polito.it

2 Department of Engineering and Applied Sciences, University of Bergamo, 24044 Dalmine, Italy; davide.disumma@unibg.it (D.d.S.); giuseppe.ruscica@unibg.it (G.R.); isabella.natali-sora@unibg.it (I.N.S.)

* Correspondence: patrizia.savi@polito.it

Received: 11 May 2020; Accepted: 14 May 2020; Published: 16 May 2020

Abstract: The partial substitution of non-renewable materials in cementitious composites with eco-friendly materials is promising not only in terms of cost reduction, but also in improving the composites' shielding properties. The water and carbon content of a commercial lignin-based biochar is analyzed with thermal gravimetric analysis. Cementitious composite samples of lignin-based biochar with 14 wt.% and 18 wt.% are realized. Good dispersion of the filler in the composites is observed by SEM analysis. The samples are fabricated in order to fit in a rectangular waveguide for measurements of the shielding effectiveness in the X-band. A shielding effectiveness of 15 dB was obtained at a frequency of 10 GHz in the case of composites with 18 wt.% biochar. Full-wave simulations are performed by fitting the measured shielding effectiveness to the simulated shielding effectiveness by varying material properties in the simulator. Analysis of the dimensional tolerances and thickness of the samples is performed with the help of full/wave simulations. Lignin-based biochar is a good candidate for partial substitution of cement in cementitious composites, as the shielding effectiveness of the composites increases substantially.

Keywords: shielding effectiveness; biochar; eco-friendly material; cementitious composites; waveguides

---

1. Introduction

The human population saw rapid growth in the past few decades. With increasing population, the demand for the construction industry increased manifold [1]. This resulted in increasing greenhouse gas emissions from cement production [2]. The substitution of non-renewable raw materials used in the construction industry with eco-friendly materials derived from waste is promising in terms of cost and environmental protection [3]. Agriculture and forestry waste is primarily burnt on field in order to reduce the cost of disposal. When converted into biochar, this waste can be used as a partial substitute to cement, resulting in a significant reduction in greenhouse gas emissions and improving the mechanical properties of concrete [4,5].

An increasing number of devices working at microwave- and millimeter-wave frequencies resulted in an overall increase in electromagnetic radiation [6,7]. Electromagnetic shields are deployed to protect sensitive devices against electromagnetic interference [8,9]. In places that are vulnerable to electromagnetic interference, shielding materials can be applied as a coating on wall surfaces [10]. A number of devices working at microwave- and millimeter-wave frequencies are used in the health sector for applications like imaging, tomography, etc. [11,12]. The X-band in particular is important for radar communications including air-traffic control, weather monitoring, maritime vessel traffic control, defense tracking, and vehicle speed detection. The use of shielding materials in buildings can be helpful in isolating equipment that is sensitive to electromagnetic interferences [13,14]. Different measurement techniques can be deployed for the determination of the shielding effectiveness of materials. The most common measurement techniques are: reverberation chamber [15], free-space measurements in an anechoic chamber [16], and coaxial and waveguide methods [17–19]. Each measurement technique requires specific sample dimensions and frequency band. The X-band is very important for applications like satellite communications and radar.

The use of carbon-based materials in epoxy composites and the analysis of their morphological and electrical properties were vastly studied [20–23]. Conventional carbon-based materials like graphene and carbon nanotubes are expensive and require complex synthesis. In recent years, the use of biochar-substituted carbon nanotubes and graphene in composites as filler was investigated [24,25]. Biochar is cost-effective as compared to other carbon-based materials. Biochar is a porous carbonaceous material produced by thermal treatment of biomass in the absence of oxygen [26]. It can be made from a number of different waste products such as agricultural waste, food waste, or sewage sludge [27]. Until recently, biochar was used for soil amendment in agriculture and landfilling applications [28]. The use of biochar in alternative applications is being studied at a vast scale, specifically for carbon sequestration, for energy storage applications [29], and in construction and building [30,31].

In this paper, lignin-based commercial biochar was used as a partial substitute to cement in composites. The water, carbon, and other residues of the biochar were studied by thermogravimetric analysis (TGA). Composites of 4 mm thickness with plain cement, 14 wt.% biochar, and 18 wt.% biochar were fabricated with specific dimensions for measurements of the shielding effectiveness inside a waveguide working in the X-band microwave frequency. The samples with 18 wt.% biochar were cured in water for seven days or 28 days. For examining the microstructural properties of the composites and dispersion of the filler in the composite matrix, SEM was adopted. Measurements of the shielding effectiveness were compared with simulated results obtained with a full-wave simulator. As expected, the shielding effectiveness increased with the increase of the percentage of filler (11 dB for 14 wt.%, and 15 dB for 18 wt.% at 10 GHz). Analyses of fabrication tolerances and sample thickness were performed with the help of a full-wave simulator.

Finally, the effect of the curing period in water on the shielding effectiveness values was analyzed for the samples with 18 wt.% biochar. The shielding effectiveness increased by approximately 5 dB in the whole frequency range for the sample cured in water for 28 days with respect to the sample cured in water for seven days.

2. Materials and Methods

2.1. Composites Preparation

The composite samples produced were with 14 wt.% and 18 wt.% biochar in Portland cement (PC). For the sake of comparison, a composite without biochar was also produced, which is referred to as plain cement composite. The biochar used to realize the samples was a commercial product provided by Carlo Erba Reagents. It was pyrolyzed in the form of powder at a temperature of 750 °C for four hours in an alumina crucible. For preparation of cementitious composites, ordinary Portland Cement (PC) (grade 52.5 R) compliant with ASTM C150 was used along with water and superplasticizer to form an adequate consistency of the paste. The percentages of water and superplasticizer used were equal to 60 wt.% and 1.8 wt.%, respectively. A mechanical mixer was used to work the mixture for a duration of 5 min. Silicon molds of adequate shape and size were then used to give the composites the required shape and dimensions.

Portland cement was blended with biochar by using a mechanical mixer for 5 min with two different percentages by weight of cement, 14% and 18%, water (60%), and superplasticizer (1.8%). Furthermore, a reference specimen was realized using only Portland cement matrix blended together with water and superplasticizer equal to 35% and 1.5%. The obtained composites were then poured into rectangular silicone molds for shielding effectiveness analysis. The silicon molds were fabricated in a three-dimensional (3D) printed master mold of specific dimensions (see Figure 1). The reusable and flexible silicone molds helped with easy extraction of composite samples once they were cured.

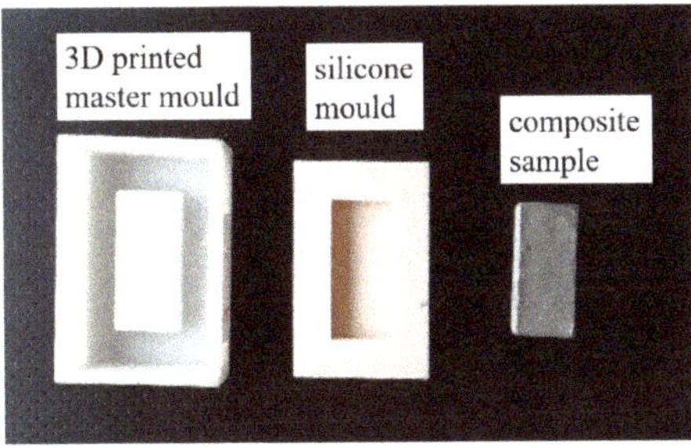

Figure 1. Three-dimensional (3D) printed master mold with silicone mold and an example of a composite sample.

Initially, the composite samples were kept at a relative humidity of 90% ± 5% for 24 h. The composites were then demolded and immersed in water at a temperature of 20 ± 2 °C. The samples were then cured in water for a period of seven days. Two different curing methodologies were used for curing of the 18 wt.% samples in water for seven days and 28 days, in order to evaluate the impact of water curing duration on the shielding effectiveness (see Table 1). In Table 1, the different steps of fabrication and measurements of the cement composites are reported.

Table 1. Fabrication and measurements of the cement composites.

Day	Plain Cement	14 wt.% (7 days)	18 wt.% (7 days)	18 wt.% (28 days)
0	Fabrication	Fabrication	Fabrication	Fabrication
1	Demolded	Demolded	Demolded	Demolded
1	Cured in water	Cured in water	Cured in water	Cured in water
7	Extracted from water	Extracted from water	Extracted from water	-
21	SE meas.* 2 weeks	SE meas.* 2 weeks	SE meas.* 2 weeks	-
28	-	-	-	Extracted from water
42	-	-	-	SE meas.* 2 weeks
70	SE meas.* 10 weeks	SE meas.* 10 weeks	SE meas.* 10 weeks	-
98	-	-	-	SE meas.* 10 weeks

* Shielding effectiveness measurement (SE meas.)

2.2. *Morphological Analysis*

Thermogravimetric analyses (TG-DTA) were carried out in air using about 20 mg of biochar heated from room temperature to 950 °C at 3 °C/min. For a morphological characterization of the cement composites, a scanning electron microscope (Hitachi S-2500C, Hitachi, Japan) was used for the analysis of the cross-section of cement composites with 18 wt.% biochar. Sections of the composite were cut and polished with measurements performed on gold-plated samples to avoid any charging effects.

2.3. *Radiofrequency Measurements*

The total shielding effectiveness can be defined as the ratio of the incident and transmitted field. It can be obtained from the measured transmission loss (S_{21}) in a waveguide as follows:

$$SE = -20 \cdot log(|S_{21}|). \tag{1}$$

The total shielding effectiveness of a material comprises dissipation loss, L_D, and mismatch loss, L_M [32].

$$SE = L_D + L_M, \tag{2}$$

where L_M can be calculated from the reflection scattering parameter as follows:

$$L_M = -10 \cdot log_{10}\left(1 - |S_{11}|^2\right), \tag{3}$$

$$L_D = -10 \cdot log_{10}\left(\frac{|S_{21}|^2}{1 - |S_{11}|^2}\right). \tag{4}$$

The scattering parameters of the composites were measured in a WR90 (Sivers IMA, Holding AB (HQ), Sweden) rectangular waveguide from 8 GHz to 12 GHz using a set-up similar to that in Reference [33]. The samples were fabricated in order to fit the rectangular waveguide cross-section (a = 22.86 mm, b = 10.16 mm). The thickness of the samples was 4 mm. The set-up is shown in Figure 2. It consisted of a two-port vector network analyzer (VNA) (Agilent E8361A: Keysight, Santa Rosa, CA 81841, USA), two coaxial cables connected to the two ports of the network analyzer, two coaxial-to-waveguide adapters, and two rectangular waveguides. Between the waveguide flanges, a spacer holding the sample was inserted. Before the measurements, a two-port calibration (short, matched load, thru) was performed. The reference planes were at the ends of the spacer.

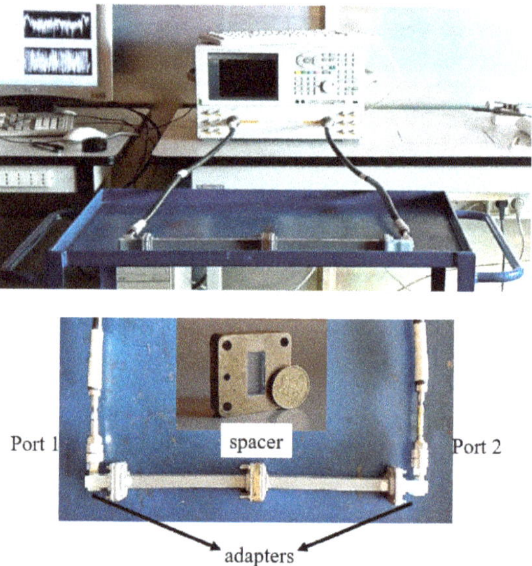

Figure 2. WR90 waveguide measurement set-up.

2.4. Finite Element Simulations

A commercial finite element modeling tool, Ansys HFSS, was used to simulate the waveguide with the composite sample as shown in Figure 3. The material properties of the composite inserted in the waveguide were chosen by fitting the simulated shielding effectiveness values to the measured shielding effectiveness values. The composite dimensions and thickness were varied to analyze the impact of fabrication tolerances and thickness on the values of shielding effectiveness.

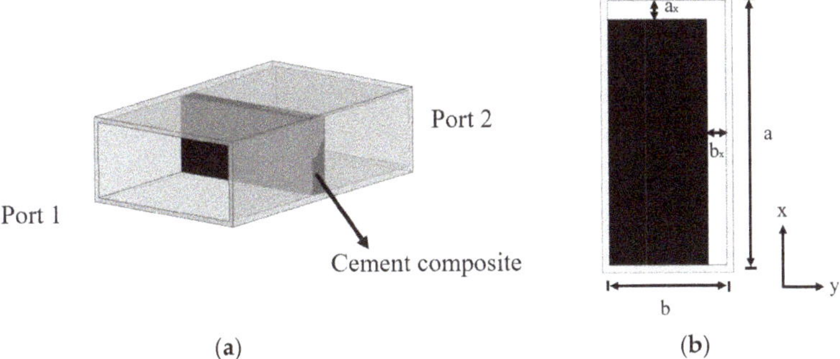

Figure 3. Geometrical configuration of the waveguide: (**a**) geometry of the simulated waveguide with composite. (**b**) Cross-section of the waveguide for the dimensional analysis.

2.5. Dimensional Tolerance Analysis

In order to take into account the dimensional tolerance of the cement composite, simulations were performed based on varying the two dimensions along the x-axis and y-axis (see Figure 3). In the case of plain cement composites, it was found that there was negligible variation of the transmission properties by varying the a_x dimension of the sample, while the impact of the variation of b_x was significant. A variation of 0.5 mm in b_x resulted in a variation of almost 1 dB in the transmission coefficient, as shown in Figure 4. It was ensured that the tolerance in the dimensions of the cement composites was below this value.

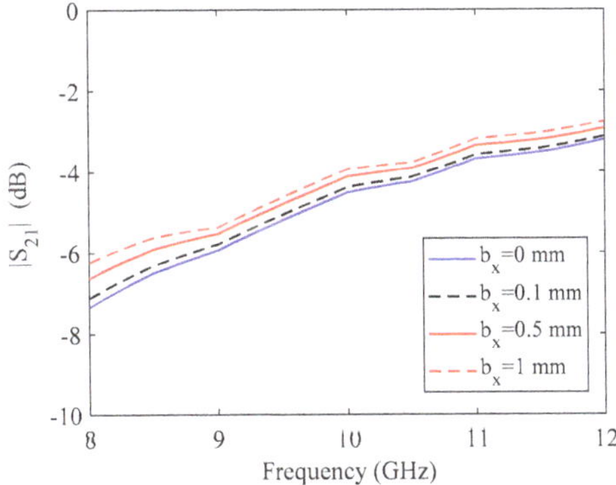

Figure 4. Analysis of fabrication tolerances of the plain cement composites.

3. Results

3.1. Biochar and Composite Characterization

The water and carbon content of the biochar was investigated by TG-DTA experiments. The TGA curve of biochar is reported in Figure 5. Below 100 °C, the weight loss was about 16%, due to the evaporation of the physically adsorbed water. From 350 °C to 500 °C, the weight loss was due to the combustion of the graphitic carbon fraction (about 74% of the total weight of the sample). At 950 °C, a residue of around 5% in weight was observed with respect to the initial amount.

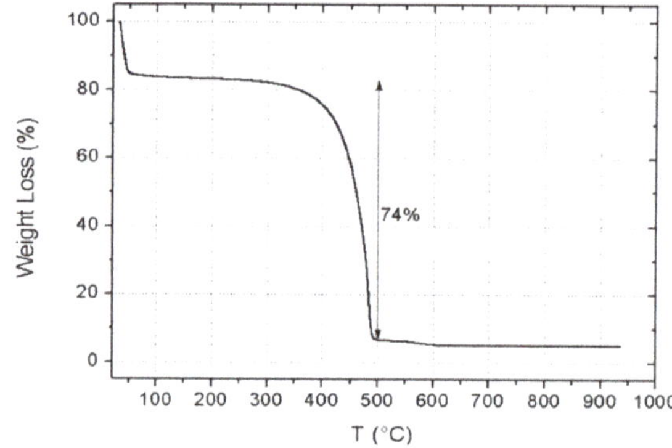

Figure 5. Thermogravimetric analysis (TGA) curve of biochar filler.

Figure 6 illustrates the SEM image of composites with the highest content of biochar (18 wt.%) recorded with secondary electrons. The black structures shown in the SEM image are the carbonaceous particles. The expected elongated structure of the particles was due to the fiber origin of the biochar. The particles showed a good dispersion in the matrix.

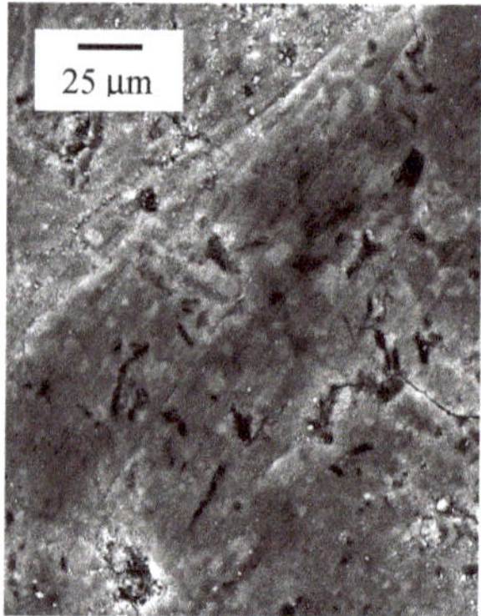

Figure 6. SEM micrograph of cement containing biochar 18% at 1000× magnification.

3.2. Shielding Effectiveness Analysis

Shielding effectiveness can be found from the measured transmission coefficient, S_{21}, in a waveguide (see Figure 2), as defined in Equation (1). The measured shielding effectiveness values of the plain cement used as a reference sample, as well as the sample with 14 wt.% and 18 wt.% filler cured in water for seven days, measured after 10 weeks, are shown in Figure 7. At the center frequency of 10 GHz, the shielding effectiveness of plain cement was almost 5 dB, which increased to 11 dB for the samples with 14 wt.% biochar. The maximum shielding effectiveness measured for the sample with 18 wt.% was around 15 dB. These results were obtained with 4-mm-thick samples. The shielding effectiveness values could be further increased by increasing the sample thickness and/or the percentage of biochar. The shielding effectiveness of the plain cement composites decreased with frequency. This behavior is similar to other cement composites [34]. The different behavior in terms of the frequency of the biochar composites with respect to plain cement composites can be attributed to the presence of entrapped water in the biochar [35].

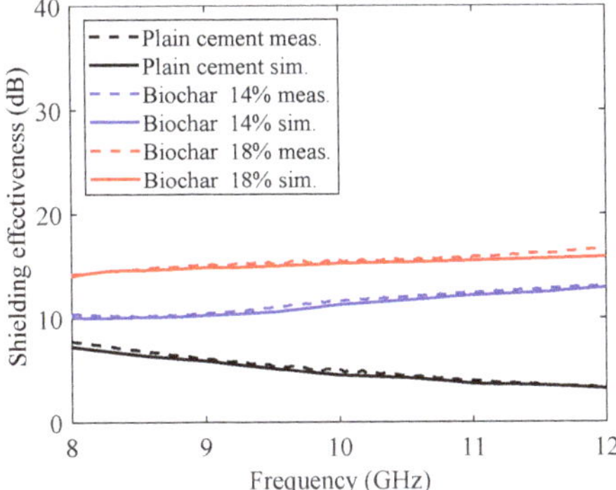

Figure 7. Measured and simulated shielding effectiveness values for plain cement, as well as the sample with 14 wt.% biochar and the sample with 18 wt.%. Samples were cured for seven days in water. Measurements were performed after 10 weeks of aging.

In Figure 7, the simulated shielding effectiveness obtained with full-wave simulations are reported (dashed lines). The values of complex permittivity were varied to fit the simulated shielding effectiveness values to the measured shielding effectiveness values, and a good correlation between the measured and simulated data was obtained.

There is a strong correlation between the curing period in water and the mechanical strength of cement composites [30]. In order to evaluate the effect of the curing period in water on the shielding effectiveness values, samples with 18 wt.% biochar cured in water for a period of seven days and 28 days were analyzed. The shielding effectiveness of the cement composite with 18 wt.% biochar cured in water for seven days and 28 days, measured after two weeks and 10 weeks, are shown in Figure 8. It can be seen that the sample cured in water for 28 days had higher shielding effectiveness when measured both after two weeks and after 10 weeks. The variation of the shielding effectiveness over time of the cement composite cured for 28 days was also higher than that cured in water for seven days. This shows that the shielding effectiveness increased due to the presence of water, whereby the loss of water from the sample over time resulted in a reduced value of the shielding effectiveness.

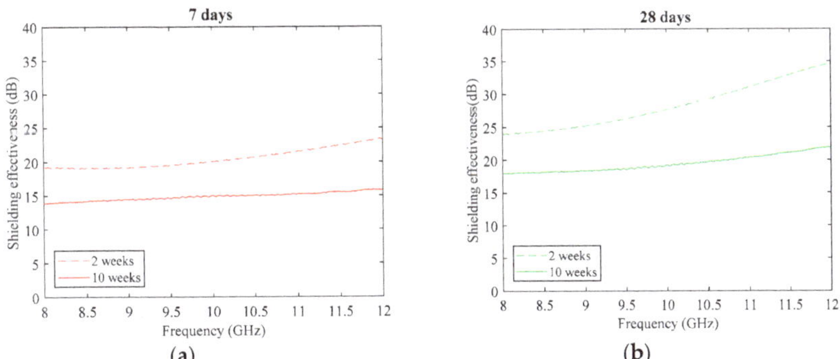

Figure 8. Measured shielding effectiveness of cement sample with 18 wt.% biochar: (**a**) cured in water for seven days; (**b**) cured in water for 28 days. Measurements were performed after two weeks and 10 weeks.

4. Discussions

In order to evaluate the impact of the presence of biochar in the cement composites on the shielding effectiveness, a comparison was performed with other studies in the literature (see Table 2).

The case considered in this comparison was the composite filled with 18 wt.% biochar cured in water for seven days and measured after 10 weeks. The thickness of the samples considered was 4 mm, which provided a shielding effectiveness value of almost 14 dB. In comparison with the literature, other reported cement samples gave higher shielding effectiveness values due to a higher value of thickness. In order to evaluate the impact of the thickness on the shielding effectiveness, simulations were performed with higher thickness values. The results are shown in Figure 9. As expected, the shielding effectiveness increased considerably upon increasing the thickness of the sample.

Table 2. Comparison with the literature.

Reference	Frequency	Measured After (days)	Thickness (mm)	Shielding Effectiveness (dB)	Materials
[34]	3 GHz	36	100	17.5	Cement
[36]	10 GHz	95	150	20	Cement
This work	10 GHz	70	4	15	Cement + 18 wt.% biochar

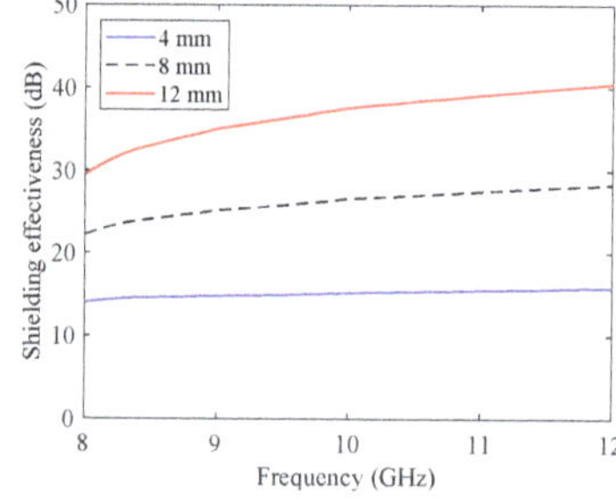

Figure 9. Simulated results for cement composites with 18 wt.% biochar with different thicknesses.

5. Conclusions

Biochar is obtained by thermal treatment of waste products. It is vastly used for soil amendment. More recently, it was used for applications such as energy storage, carbon sequestration, and construction. The effect of a commercial biochar on the shielding properties of cement composites was investigated in the X-band. The conclusions drawn based on the results presented can be extended to other microwave frequencies. Cementitious composites with ordinary Portland Cement (PC) were prepared without biochar and with biochar as filler (14 wt.% and 18 wt.%). Samples were prepared in order to fit a WR90 waveguide (8–12 GHz). With the help of a full-wave simulator, the fabrication tolerances of the samples were analyzed. A variation of ±0.5 mm resulted in a change of the shielding effectiveness of ±1 dB. Shielding effectiveness can be obtained from the measurements of scattering parameters. Samples with 14 wt.% and 18 wt.% biochar as filler were cured in water for seven days. As expected, the shielding effectiveness increased with the increase in the percentage of filler (11 dB for 14 wt.%, and 15 dB for 18 wt.% at 10 GHz). In order to evaluate the effect of the curing period in water on the shielding effectiveness, different curing periods were analyzed. Samples with 18 wt.% biochar were cured in water for a period of seven days and 28 days. The shielding effectiveness increased by approximately 5 dB in the whole frequency range for the samples cured in water for 28 days as compared to samples cured in water for seven days.

Author Contributions: Composite fabrication, D.d.S. and G.R.; waveguide measurements and discussion of the shielding effectiveness, D.d.S., M.Y., and P.S.; microstructure characterization and TGA, I.N.S.; full-wave simulations, M.Y.; original draft preparation, M.Y. and P.S.; writing—review and editing, M.Y., P.S., and I.N.S.; supervision, P.S.; conceptualization, M.Y., P.S., and I.N.S., funding acquisition, G.R. All authors read and agreed to the published version of the manuscript.

Funding: This research received no external funding.

Acknowledgments: The authors would like to thank Renato Pelosato for TGA measurements.

Conflicts of Interest: The authors declare no conflict of interest.

References

1. Klee, H. Briefing: The Cement Sustainability Initiative. *Proc. Inst. Civ. Eng.—Eng. Sustain.* **2004**, *157*, 9–11. [CrossRef]
2. Oh, D.-Y.; Noguchi, T.; Kitagaki, R.; Park, W.-J. CO_2 emission reduction by reuse of building material waste in the Japanese cement industry. *Renew. Sust. Energ. Rev.* **2014**, *38*, 796–810. [CrossRef]
3. Balasubramanian, J.; Gopal, E.; Periakaruppan, P. Strength and microstructure of mortar with sand substitutes. *Graevinar* **2016**, *68*, 29–37.
4. Van der Lugt, P.; Van den Dobbelsteen, A.A.J.F.; Janssen, J.J.A. An environmental, economic and practical assessment of bamboo as a building material for supporting structures. *Constr. Build. Mater.* **2006**, *20*, 648–656. [CrossRef]
5. Klapiszewski, Ł.; Klapiszewska, I.; Ślosarczyk, A.; Jesionowski, T. Lignin-Based Hybrid Admixtures and their Role. *Cem. Compos. Fabr. Mol.* **2019**, *24*, 3544.
6. Research and Markets. Available online: https://www.researchandmarkets.com/research/6mzxvg/microwave_devices (accessed on 25 February 2020).
7. Delhi, N.; Behari, G. Electromagnetic pollution-the causes and concerns. In Proceedings of the International Conference of Electromagnetic Interference and Compatibility, Bangalore, India, 23 February 2002.
8. McKerchar, W.D. Electromagnetic Compatibility of High Density Wiring Installations by Design or Retrofit. *IEEE Trans Elm. Comp.* **1965**, *7*, 1–9. [CrossRef]
9. Wang, Y.; Gordon, S.; Baum, T.; Su, Z. Multifunctional Stretchable Conductive Woven Fabric Containing Metal Wire with Durable Structural Stability and Electromagnetic Shielding in the X-Band. *Polymers* **2020**, *12*, 399. [CrossRef]
10. Lee, H.S.; Park, J.H.; Singh, J.K.; Choi, H.J.; Mandal, S.; Jang, J.M.; Yang, H.M. Electromagnetic Shielding Performance of Carbon Black Mixed Concrete with Zn–Al Metal Thermal Spray Coating. *Materials* **2020**, *13*, 895. [CrossRef]
11. Mojtaba, A.; Maryam, I.; Saripan, A.; Iqbal, M.; Hasan, W.Z.W. Three dimensions localization of tumors in confocal microwave imaging for breast cancer detection. *Microw. Opt. Technol. Lett.* **2015**, *57*, 2917–2929.
12. Peng, K.C.; Lin, C.C.; Li, C.F.; Hung, C.Y.; Hsieh, Y.S.; Chen, C.C. Compact X-Band Vector Network Analyzer for Microwave Image Sensing. *IEEE Sens. J.* **2019**, *9*, 3304–3313. [CrossRef]
13. Hanada, E.; Watanabe, Y.; Antoku, Y.; Kenjo, Y.; Nutahara, H.; Nose, Y. Hospital construction materials: Poor shielding capacity with respect to signals transmitted by mobile telephones. *Biomed. Instrum. Technol.* **1998**, *32*, 489–496. [PubMed]
14. Khushnood, R.A.; Ahmad, S.; Savi, P.; Tulliani, J.M.; Giorcelli, M.; Ferro, G.A. Improvement in electromagnetic interference shielding effectiveness of cement composites using carbonaceous nano/micro inerts. *Constr. Build. Mater.* **2015**, *85*, 208–216. [CrossRef]
15. Holloway, C.L.; Hill, D.A.; Ladbury, J.; Koepke, G.; Garzia, R. Shielding effectiveness measurements of materials using nested reverberation chambers. *IEEE Trans. Electromagn. Compat.* **2003**, *45*, 350–356. [CrossRef]
16. Jung, M.; Lee, Y.-S.; Hong, S.-G. Effect of Incident Area Size on Estimation of EMI Shielding Effectiveness for Ultra-High Performance Concrete With Carbon Nanotubes. *IEEE Access* **2019**, *17*, 183106–183117. [CrossRef]
17. Tamburrano, A.; Desideri, D.; Maschio, A.; Sarto, S. Coaxial Waveguide Methods for Shielding Effectiveness Measurement of Planar Materials Up to 18 GHz. *IEEE Trans. Electromagn. Compat.* **2014**, *56*, 1386–1395. [CrossRef]
18. Valente, R.; Ruijter, C.D.; Vlasveld, D.; Zwaag, S.V.D.; Groen, P. Setup for EMI Shielding Effectiveness Tests of Electrically Conductive Polymer Composites at Frequencies up to 3.0 GHz. *IEEE Access* **2017**, *5*, 16665–16675. [CrossRef]
19. Rudd, M.; Baum, T.C.; Ghorbani, K. Determining High-Frequency Conductivity Based on Shielding Effectiveness Measurement Using Rectangular Waveguides. *IEEE Trans. Instrum. Meas.* **2020**, *69*, 155–162. [CrossRef]
20. Gupta, S.; Tai, N.H. Carbon materials and their composites for electromagnetic interference shielding effectiveness in X-band. *Carbon* **2019**, *152*, 159–187. [CrossRef]

21. Giorcelli, M.; Savi, P.; Yasir, M.; Miscuglio, M.; Yahya, M.H.; Tagliaferro, A. Investigation of epoxy resin/multiwalled carbon nanotube nanocomposites behavior at low frequency. *J. Mater. Res.* **2014**, *30*, 101–107. [CrossRef]
22. Savi, P.; Yasir, M.; Giorcelli, M.; Tagliaferro, A. The effect of carbon nanotubes concentration on complex permittivity of nanocomposites. *Prog. Electromagn. Res. M* **2017**, *55*, 203–209. [CrossRef]
23. Khan, A.; Savi, P.; Quaranta, S.; Rovere, M.; Giorcelli, M.; Tagliaferro, A.; Rosso, C.; Jia, C.Q. Low-Cost Carbon Fillers to Improve Mechanical Properties and Conductivity of Epoxy Composites. *Polymers* **2017**, *9*, 642. [CrossRef] [PubMed]
24. Peterson, S.C. Evaluating corn starch and corn stover biochar as renewable filler in carboxylated styrene butadiene rubber composites. *J. Elastomers Plast.* **2011**, *44*, 43–54. [CrossRef]
25. Giorcelli, M.; Savi, P.; Khan, A.; Tagliaferro, A. Analysis of biochar with different pyrolysis temperatures used as filler in epoxy resin composites. *Biomass Bioenergy* **2019**, *122*, 466–471. [CrossRef]
26. Bridgwater, A.V. Review of fast pyrolysis of biomass and product upgrading. *Biomass Bioenergy* **2012**, *38*, 68–94. [CrossRef]
27. Savi, P.; Yasir, M.; Bartoli, M.; Giorcelli, M.; Longo, M. Electrical and Microwave Characterization of Thermal Annealed Sewage Sludge Derived Biochar Composites. *Appl. Sci.* **2020**, *10*, 1334. [CrossRef]
28. Ding, Y.; Liu, Y.; Liu, S.; Li, Z.; Tan, X.; Huang, X.; Zeng, G.; Zhou, L.; Zheng, B. Biochar to improve soil fertility. A review. *Agron. Sustain. Dev.* **2016**, *36*, 36. [CrossRef]
29. Ngan, A.; Jia, C.Q.; Tong, S.-T. Production, Characterization and Alternative Applications of Biochar. In *Production of Materials from Sustainable Biomass Resources*; Springer: Singapore, 2019; pp. 117–151.
30. Gupta, S.; Kua, H.W.; Low, C.Y. Use of biochar as carbon sequestering additive in cement mortar. *Cem. Concr. Compos.* **2018**, *87*, 110–129. [CrossRef]
31. Gupta, S.; Kua, H.W.; Pang, S.D. Effect of biochar on mechanical and permeability properties of concrete exposed to elevated temperature. *Constr. Build. Mater.* **2020**, *234*, 117338. [CrossRef]
32. Savi, P.; Yasir, M. Waveguide measurements of biochar derived from sewage sludge. *Electron. Lett.* **2020**, *56*, 335–337. [CrossRef]
33. Savi, P.; Cirielli, D.; di Summa, D.; Ruscica, G.; Sora, I.N. Analysis of shielding effectiveness of cement composites filled with pyrolyzed biochar. In Proceedings of the 2019 IEEE 5th International forum on Research and Technology for Society and Industry (RTSI), Florence, Italy, 9–12 September 2019; pp. 376–379.
34. Donnell, K.M.; Zoughi, R.; Kurtis, K.E. Demonstration of Microwave Method for Detection of Alkali–Silica Reaction (ASR) Gel in Cement-Based Materials. *Cem. Concr. Res.* **2013**, *44*, 1–7. [CrossRef]
35. Mrad, R.; Chehab, G. Mechanical and Microstructure Properties of Biochar-Based Mortar: An Internal Curing Agent for PCC. *Sustainability* **2019**, *11*, 2491. [CrossRef]
36. Kharkovsky, S.N.; Akay, M.F.; Hasar, U.C.; Atis, C.D. Measurement and monitoring of microwave reflection and transmission properties of cement-based specimens. *IEEE Trans. Instrum. Meas.* **2002**, *51*, 1210–1218. [CrossRef]

Article

Synthesis and Characterization of Polyaniline-Based Composites for Electromagnetic Compatibility of Electronic Devices

Kamil G. Gareev *, Vladislava S. Bagrets, Vladimir A. Golubkov, Maria G. Ivanitsa, Ivan K. Khmelnitskiy, Victor V. Luchinin, Olga N. Mikhailova and Dmitriy O. Testov

Department of Micro and Nanoelectronics, Saint Petersburg Electrotechnical University "LETI", 197376 Saint Petersburg, Russia; bagretsvlada@mail.ru (V.S.B.); vavantess@mail.ru (V.A.G.); marie.ivanitsa@gmail.com (M.G.I.); khmelnitskiy@gmail.com (I.K.K.); cmid_leti@mail.ru (V.V.L.); novaja.tasamaya@yandex.ru (O.N.M.); dtestov@bk.ru (D.O.T.)

* Correspondence: kggareev@etu.ru

Received: 13 April 2020; Accepted: 26 April 2020; Published: 29 April 2020

Abstract: Polyaniline-based composites designed to ensure the electromagnetic compatibility of electronic devices were obtained. The surface morphologies of the obtained films were studied using optical and electron microscopy. The electrical resistivity of polyaniline (PANI) films were measured at various thicknesses. For films of various compositions and various thicknesses, the frequency dependencies of the complex dielectric permittivity, in the range of 100–2000 kHz, as well as the electromagnetic radiation (EMR) absorption coefficient in the frequency range 0.05–2 GHz were obtained. It was found that flexible gelatin-PANI composite films with a thickness of 200–400 μm, a bending radius of about 5 cm, and a real part of complex permittivity of not more than 10 provide an EMR absorption coefficient of up to 5 dB without introducing additional EMR absorbing or reflecting fillers. The resulting gelatin-PANI composite films do not possess a through electrical conductivity and can be applied directly to the surface of protected printed circuit boards.

Keywords: electromagnetic compatibility; polyaniline; gelatin; composite; microwave absorption; dielectric permittivity; electrical conductivity

1. Introduction

Electromagnetic compatibility (EMC) of electronic devices is ensured by applying materials shielding from electromagnetic radiation (EMR) due to EMR absorption and reflection [1]. Ferrite-based composites with organic binders [2] are widely used materials for EMR absorption. Their serious disadvantage is a high mass density, which limits the application of such materials in electronics. The use of composite materials containing conductive polymers and components with high dielectric and magnetic losses can be considered as the most promising way to ensure EMC. Conductive polymer polyaniline (PANI) is used as an additional component with dielectric losses in a wide variety of EMC composites. A high EMR shielding efficiency, in a wide frequency range, has been experimentally proven for composites of PANI using silver nanowires [3], particles of antimony oxide [4], graphene [5–7], carbon nanotubes [8,9], Ti_3SiC_2 [10], MoS_2 [11], CuO [12], bamboo fibers [13], bagasse [14], MXene, and other fillers [15]. The EMR shielding properties of composites of PANI with magnetic nanoparticles of cobalt and iron–nickel alloy [16] and ferrites [17] have also been shown.

PANI was applied as a component of composites for EMC. PANI/polyacrylate composite coatings demonstrated the efficiency of electromagnetic-interference shielding at 38–60 dB in a frequency range of 100 kHz–10 GHz [18]. PANI nanofibers coatings showed an efficiency of electromagnetic-interference shielding of up to 63 dB in that same frequency range [19]. Films of conductive polymeric blends of polystyrene and PANI provided a shielding efficiency of up to 45 dB in a frequency range of 9–18 GHz [20]. PANI provided a low-percolation threshold (0.58 wt.%), high-electrical conductivity, and high-shielding efficiency of composites using carbon nanotubes [21]. Composites based on PANI and MXene possess excellent thermal and EMR shielding efficiency characteristics; thus, they can be applied in for use in mobile phones, military utensils, heat-emitting electronic devices, automobiles, and radars [22]. Even though the conductivity of PANI is not as high as that of metals, it is considerably high when compared to the conductivity of other polymers. Additionally, PANI has many advantages, including ease of fabrication, low cost, and the ability to be switched between electrically conducting and insulating statuses [15].

The efficiency of PANI-containing composites without additional EMR absorbing or reflecting fillers is not widely described in the literature, but these composites may be useful in order to simplify synthesis techniques and to reduce the costs of EMC materials. At the same time, composites of gelatin–PANI, which are potentially useful for various fields, including electronics [23], are only characterized as materials for tissue engineering [24,25], targeted drug delivery [26], and other biomedical applications [27].

The aim of the current work was to obtain PANI films and gelatin–PANI composite films and to study their potential applications for the EMC of electronic devices.

2. Materials and Methods

2.1. Synthesis of PANI Films

In order to obtain a PANI dispersion, a solution containing 0.228 g of $(NH_4)_2S_2O_8$ (Merck, USA) and 1 mL of distilled water was added to a solution containing 0.255 mL of aniline (Merck, USA), 1 mL of 10 M HCl (Merck, USA), and 7 mL of distilled water. The reaction mixture was kept in a water bath at 20 °C for 24 h. The resulting suspension was dark green, typical for the protonated emeraldine form of PANI, which has the highest stability and the lowest resistivity according to the results of other research groups [28]. Then, the suspension was filtered, washed several times with distilled water, refiltered, and dried to obtain a solid PANI, from which a suspension of a given concentration was prepared. PANI films were fabricated with the help of irrigation on polyethylene substrates with a size of 44 × 44 mm, followed by being dried in an oven at 40 °C. The thickness of films ranged from 50 to 200 μm.

2.2. Synthesis of Gelatin–PANI Composites

Obtaining samples of gelatin–PANI composites (for example, 12%-PANI-GEL with 12 mas.% PANI) was carried out as follows: 6.5 g of gelatin P-11 (GOST 11293-89, Vekton, Saint Petersburg, Russia) was added to 50 mL of distilled water in a glass flask. The produced suspension was being stirred using a magnetic stirrer at a temperature of 60 °C for 2 h. An aqueous suspension of polyaniline was prepared in a conical tube by adding a volume of of dry PANI (obtained in accordance with Section 2.1) of 0.28 g to 5.3 mL of distilled water. Further on, the PANI suspension was treated with an ultrasonic homogenizer (Sonopuls HD 4100 (Bandelin, Berlin, Germany)) for 5 min, directly before being mixed with a gelatin solution. The final suspension was poured into a Petri dish for solidification at room temperature.

The samples, 3%-PANI-GEL, 22%-PANI-GEL, 36%-PANI-GEL, 46%-PANI-GEL, 53%-PANI-GEL, and 59%-PANI-GEL with 3, 22, 36, 46, 53, and 59 mas.% PANI, respectively, were obtained in a similar manner. Each synthesis experiment was repeated five times. The maximum content of PANI was due to the need to obtain a stable suspension for sample preparation.

To provide electrophysical measurements, a sample with a size of 44 × 44 mm and a thickness of 200–400 μm was cut out of the obtained flexible film. As a reference sample, a gelatin-based film with no addition of conductive polymer was obtained.

2.3. Characterization Techniques

The surface morphology of the obtained films was characterized by optical and scanning electron microscopy (SEM) using a digital optical microscope (KH-7700 (Hirox, Tokyo, Japan)) and a two-beam scanning electron microscope (Helios Nanolab 400 (FEI, Hillsboro, OR, USA)).

The frequency dependencies of complex dielectric permittivity were measured in accordance with ASTM D150, using an Agilent E4980A high-precision LCR meter (Agilent Technologies, Inc., Santa Clara, CA, USA) and a measuring cell developed at Saint Petersburg Electrotechnical University "LETI" [29].

To measure the EMR absorption coefficient, we used a method based on a coplanar waveguide (CPW) with an impedance of 50 Ω, assigned to the frequency range of 50–2000 MHz, and a vector network analyzer ZVB-20 (Rohde&Schwarz, Munich, Germany). The design of the CPW was developed at Saint Petersburg Electrotechnical University "LETI" [29]. When measuring, the test sample was placed on the surface of the CPW.

The EMR absorption coefficient L (dB) was calculated using the following formula [30]:

$$L = -10\log_{10}\left(|S_{21}|^2 + |S_{11}|^2\right),$$

where $|S_{21}|$ and $|S_{11}|$ are the modules of complex transmission and reflection coefficients, respectively. The absorption coefficient's value, 10 dB, was equivalent to 90% power loss of transmitted EMR.

A 4200-SCS semiconductor characterization system (Keithley, Solon, OH, USA) and an M150 probe station (Cascade Microtech, Beaverton, OR, USA) were used to measure the surface electrical resistivity of samples using the four-probe method. In order to calculate the electrical resistivity of the samples, their thicknesses were measured using a micrometer with an accuracy of 5 μm [29]. Each sample was measured five times.

3. Results and Discussion

3.1. Surface Morphology of the Samples

The views of the obtained PANI film sample on a polyethylene substrate, the sample of gelatin film, and the sample of the gelatin–PANI composite film are given in Figure 1. The minimal binding radius of the gelatin-PANI composite film with a thickness of 200–400 μm was about 5 cm.

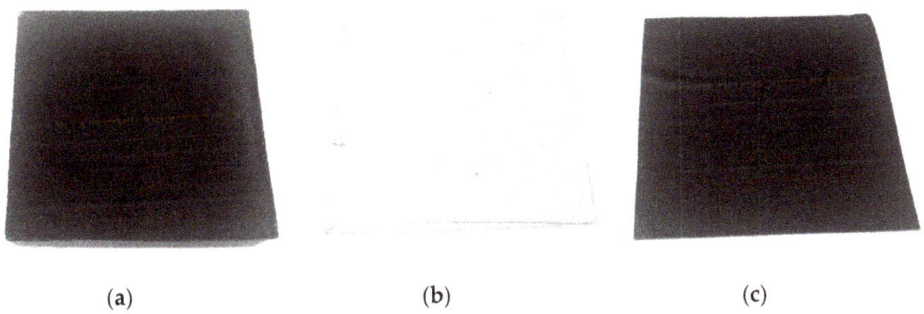

(a) (b) (c)

Figure 1. The view of the obtained samples: (**a**) the polyaniline (PANI) film on a polyethylene substrate; (**b**) the gelatin film; and (**c**) the gelatin-PANI composite film.

SEM images of the PANI film on a polyethylene substrate are given in Figure 2. As can be seen from the images, the film consists of PANI grains of 100–1000 nm, which presumably may be due to the fact that the film was directly obtained from an aniline solution with subsequent polymerization.

(a) (b)

Figure 2. SEM images of the PANI film on a polyethylene substrate: (**a**) low magnification and (**b**) high magnification.

The optical microscopy images of PANI film on a polyethylene substrate and gelatin–PANI samples are given in Figure 3.

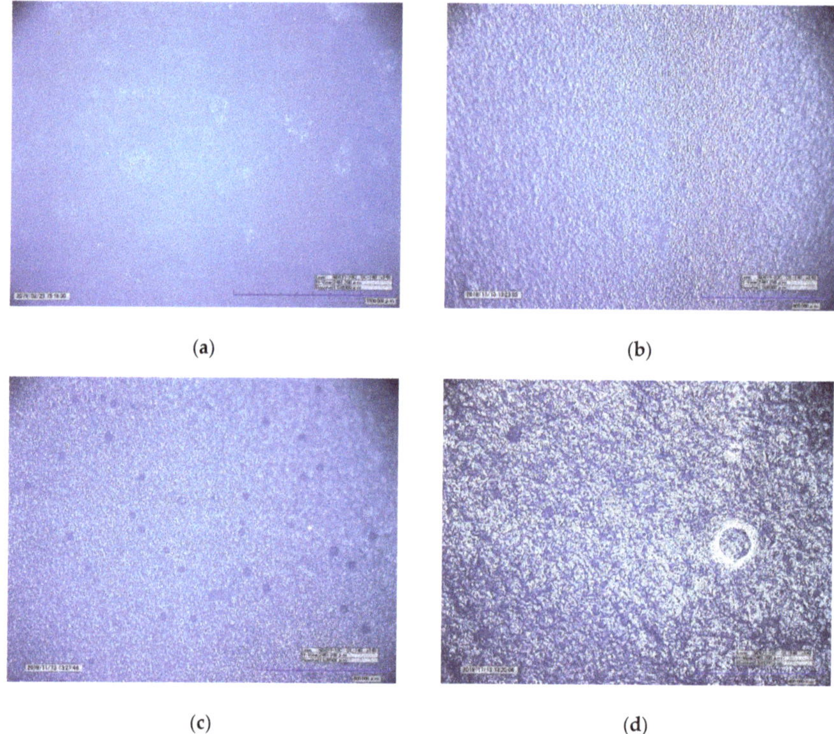

(a) (b)

(c) (d)

Figure 3. *Cont.*

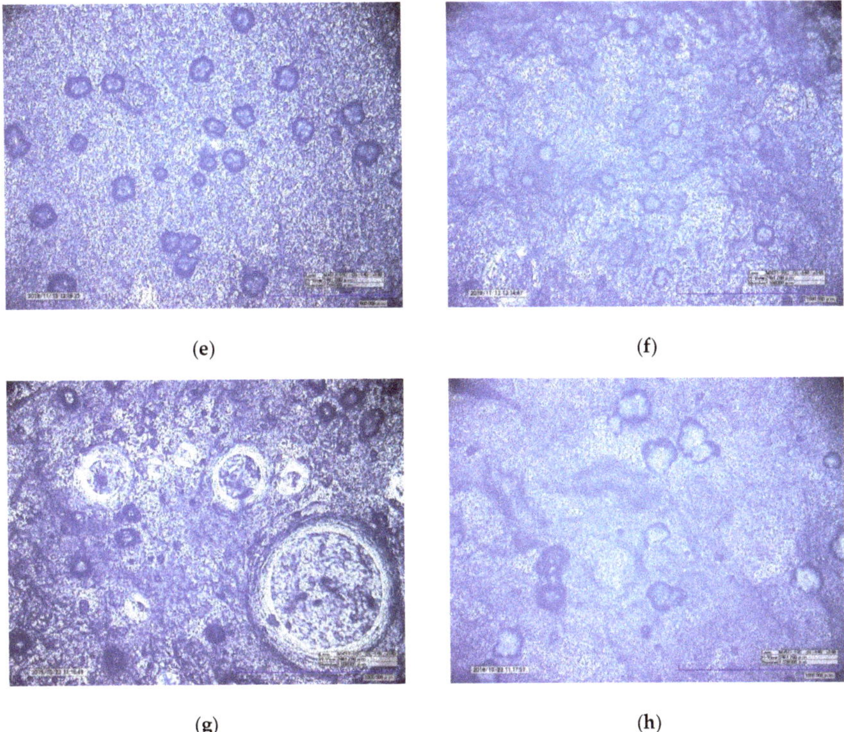

Figure 3. Optical microscopy images of the samples: (**a**) the PANI film on a polyethylene substrate; (**b**) 3%-PANI-GEL; (**c**) 12%-PANI-GEL; (**d**) 22%-PANI-GEL; (**e**) 36%-PANI-GEL; (**f**) 46%-PANI-GEL; (**g**) 53%-PANI-GEL; and (**h**) 59%-PANI-GEL.

As shown in the obtained images, an increase in PANI content in the composition provokes the formation of larger particles of conductive polymer. A further increase in the content of PANI is possible with changes in synthesis technology.

3.2. Electrophysical Characteristics

The results of the electrical resistivity of the PANI films on a polyethylene substrate are shown in Figure 4.

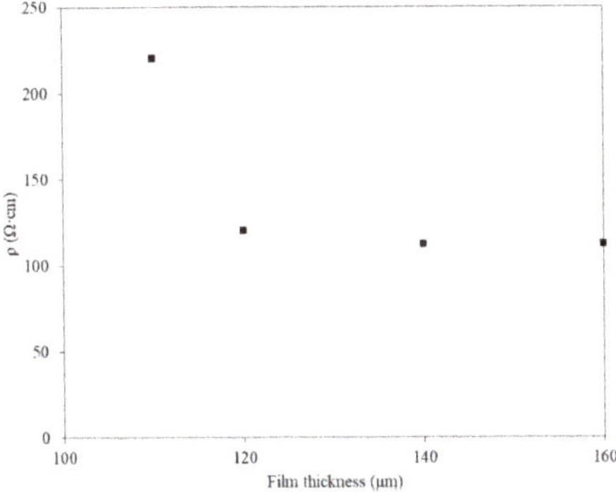

Figure 4. Thickness dependence of the electrical resistivity of the PANI films on a polyethylene substrate.

The electrical resistivity of PANI films decreases, whereas the film thickness increases and stops its decrease at a thickness of about 100–120 μm. The effect is similar to that described in [31] and may be due to the compaction of the structure of the PANI film on a nonconductive polyethylene substrate [32]. The value of the surface electrical resistivity of the PANI film may be increased by introducing additional dielectric components, or by selecting the PANI content in the gelatin–PANI composite in order to achieve an impedance matching with free space ($Z = 377\ \Omega$).

The frequency dependencies of the real and imaginary parts of the complex dielectric permittivity of gelatin–PANI composite samples are shown in Figure 5.

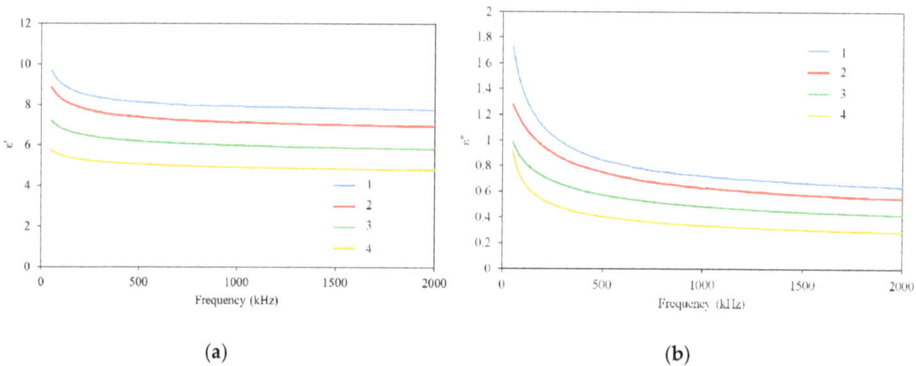

Figure 5. Frequency dependence of the complex dielectric permittivity of the gelatin–PANI composite films: (**a**) real part of dielectric permittivity; (**b**) imaginary part of dielectric permittivity; 1—gelatin; 2—22%-PANI-GEL; 3—36%-PANI-GEL; and 4—59%-PANI-GEL.

As can be seen from Figure 5, the maximum value of the real part of complex dielectric permittivity does not exceed 10; therefore, considering the absence of through electrical conductivity, the composite is close to a dielectric material and its EMR reflection coefficient can be low. An increase in PANI content leads to a decrease in real and imaginary parts of complex dielectric permittivity.

The effect may be due to several different possible reasons. The first is the fact that the dielectric permittivity of replaced gelatin is much higher in comparison with the dielectric permittivity of the gelatin–PANI composites [33]. The second reason is the structural features of the composite, which cause the appearance of electrical moments due to electrically isolated PANI particles, which can be represented as elementary electric dipoles. An increase in the size of PANI grains observed experimentally via optical microscopy leads to a decrease in the number of electric dipoles, a decrease in the electric moment of a unit volume, and, accordingly, a decrease in dielectric permittivity. Composites based on various conductive polymers can have ferroelectric properties and be characterized by a high-dielectric constant, which can be controlled by an external electric field or temperature [34,35]. The third possible reason is not due to physical processes within the composite and is due to the high inhomogeneity of gelatin-PANI composite film with a high content of PANI, which affects the measured values of complex dielectric permittivity due to an increased air gap between the measuring electrodes and the film surfaces.

3.3. EMR Absorption Measurement

The frequency dependencies of the EMR absorption coefficient for samples of PANI films of various thicknesses are shown in Figure 6.

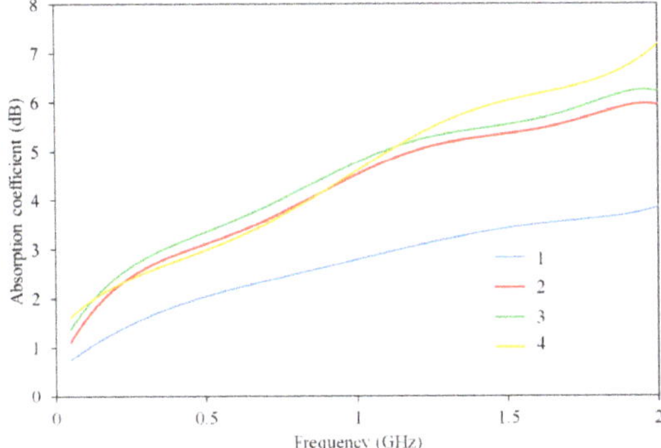

Figure 6. Frequency dependence of the electromagnetic radiation (EMR) absorption coefficient of the PANI films on a polyethylene substrate: 1—110 μm, 2—120 μm, 3—140 μm, and 4—160 μm.

The value of the EMR absorption coefficient reached 6–7 dB for PANI films with a thickness of over 120 μm, and changed insignificantly, while the thickness increased further. These results correlate with the data on the films' electrical resistivity measurements; hence, the assumption of the prevalent role of dielectric loss in PANI can be confirmed. At the same time, an application of continuous PANI films without a dielectric binder is restricted by a mismatch of their impedance with free space impedance and, consequently, by a high EMR reflection coefficient. For this reason, in our opinion, an estimation of the EMR absorption properties of gelatin–PANI composite films may be much more useful. The frequency dependence of the EMR absorption coefficient of gelatin–PANI composite films is shown in Figure 7.

As can be seen, the EMR absorption coefficient grows monotonously and reaches 4–5 dB, while the PANI content increases to 50 mas.%. In accordance with the optical microscopy and complex dielectric permittivity measurement data, the observed effect may be explained by two possible reasons. The first one is the change in film surface morphology, which, due to its high inhomogeneity, leads to an increase in the air gap between the CPW and film surfaces. Therefore, a further increase in the measured value of the EMR absorption coefficient becomes impossible. The second possible reason is a decrease in dielectric losses with an increase in PANI content. If the EMR saturation value maximum's origin is a technological limitation, it can be overcome by making changes to the synthesis technique.

The physical mechanisms of EMR absorption should be similar to those proposed in [20] since it considered composite films of the same thicknesses (250 μm) and with a similar PANI content (40 mas.%) in a dielectric binder, polystyrene. However, a direct comparison of the values of the absorption coefficients is not entirely correct due to the different frequency ranges of measurements (0.05–2 GHz and 9–18 GHz) and different measurement techniques (air-coaxial line and CPW). Nevertheless, the authors believe that the gelatin–PANI composite films will demonstrate similar or higher values of EMR absorption coefficient at frequencies over 8 GHz because, unlike polystyrene, gelatin binders possess intrinsic dielectric losses [29].

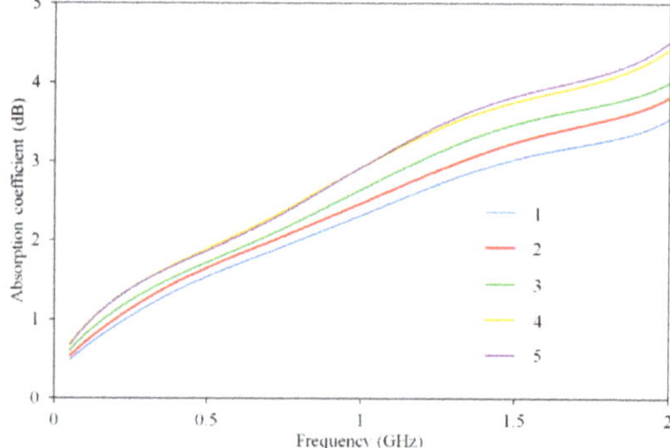

Figure 7. Frequency dependence of the EMR absorption coefficient of the gelatin-PANI composite films: 1—2%-PANI-GEL; 2—36%-PANI-GEL; 3—46%-PANI-GEL; 4—53%-PANI-GEL; and 5—22%-PANI-GEL.

4. Conclusions

As a result of the study, it was found that PANI-based composite films with a film thickness of about 200–400 μm and a minimum bending radius of about 5 cm are capable of providing an EMR absorption coefficient of up to 5 dB in a frequency range 0.05–2 GHz. The possibility of using the obtained composite films without introducing additional absorbing fillers (carbon nanotubes, graphene, magnetic nanoparticles, etc.) into the composition allows to offer a cheap and technologically advanced material to ensure the EMC of electronic devices designed for standard application conditions. Moreover, due to the low value of the real part of complex dielectric permittivity (not more than 10) and the absence of through electrical conductivity, as well as the low temperature of production (not higher than 60 °C), gelatin–PANI composite films can be applied directly to printed circuit boards to form a continuous conformal protective layer on their surfaces.

Author Contributions: Conceptualization, I.K.K. and V.V.L.; methodology, I.K.K. and K.G.G.; formal analysis, I.K.K. and K.G.G.; investigation, K.G.G., V.S.B., V.A.G., M.G.I., O.N.M. and D.O.T.; data curation, K.G.G.; writing—original draft preparation, K.G.G.; writing—review and editing, I.K.K. and V.V.L.; visualization, K.G.G.; supervision, V.V.L.; project administration, I.K.K.; funding acquisition, I.K.K. and V.V.L. All authors have read and agreed to the published version of the manuscript.

Funding: This research was funded by Russian Science Foundation, grant number 16-19-00107.

Conflicts of Interest: The authors declare no conflicts of interest.

References

1. Ott, H.W. *Electromagnetic Compatibility Engineering*; John Wiley & Sons: Hoboken, NJ, USA, 2009; pp. 243–245. ISBN 978-0-470-18930-6.
2. Mouritz, A.P. *Introduction to Aerospace Materials*; Woodhead Publishing: Oxford, UK, 2012; pp. 296–298, ISBN 978-0-85709-515-2.
3. Fang, F.; Li, Y.-Q.; Xiao, H.-M.; Hu, H.; Fu, S.-Y. Layer-structured silver nanowire/polyaniline composite film as high performance X-band EMI shielding material. *J. Mater. Chem. C* **2016**, *19*, 4193–4203. [CrossRef]
4. Faisal, M.; Khasim, S. Polyaniline–antimony oxide composites for effective broadband EMI shielding. *Iran Polym. J.* **2013**, *22*, 473–480. [CrossRef]
5. Khasim, S. Polyaniline-Graphene nanoplatelet composite films with improved conductivity for high performance X-band microwave shielding applications. *Res. Phys.* **2019**, *12*, 1073–1081. [CrossRef]
6. Shakir, M.F.; Khan, A.N.; Khan, R.; Javed, S.; Tariq, A.; Azeem, M.; Riaz, A.; Shafqat, A.; Cheema, H.M.; Akram, M.A.; et al. EMI shielding properties of polymer blends with inclusion of graphene nano platelets. *Res. Phys.* **2019**, *14*, 102365. [CrossRef]
7. Cheng, K.; Li, H.; Zhu, M.; Qiu, H.; Yang, J. In situ polymerization of graphene-polyaniline@polyimide composite films with high EMI shielding and electrical properties. *RSC Adv.* **2020**, *10*, 2368–2377. [CrossRef]

8. Kumar, A.; Kumar, V.; Kumar, M.; Awasthi, K. Synthesis and Characterization of Hybrid PANI/MWCNT Nanocomposites for EMI Applications. *Polym. Compos.* **2018**, *39*, 3858–3868. [CrossRef]
9. Li, H.; Lu, X.; Yuan, D.; Sun, J.; Erden, F.; Wang, F.; He, C. Lightweight flexible carbon nanotube/polyaniline films with outstanding EMI shielding properties. *J. Mater. Chem. C* **2017**, *5*, 8694–8698. [CrossRef]
10. Shi, S.; Zhang, L.; Li, J. Complex permittivity and electromagnetic interference shielding properties of Ti_3SiC_2/polyaniline composites. *J. Mater. Sci.* **2009**, *44*, 945–948. [CrossRef]
11. Saboor, A.; Khalid, S.M.; Jan, R.; Khan, A.N.; Zia, T.; Farooq, M.U.; Afridi, S.; Sadiq, M.; Arif, M. PS/PANI/MoS_2 Hybrid Polymer Composites with High Dielectric Behavior and Electrical Conductivity for EMI Shielding Effectiveness. *Materials* **2019**, *12*, 2690. [CrossRef]
12. Rahul, D.S.; Pais, T.P.M.; Sharath, N.; Ali, S.A.; Faisal, M. Polyaniline-copper oxide composite: A high performance shield against electromagnetic pollution. *AIP Conf. Proc.* **2015**, *1665*, 140021. [CrossRef]
13. Yang, Z.; Zhag, Y.; Wen, B. Enhanced electromagnetic interference shielding capability in bamboo fiber@polyaniline composites through microwave reflection cavity design. *Compos. Sci. Technol.* **2019**, *178*, 41–49. [CrossRef]
14. Zhang, Y.; Qiu, M.; Yu, Y.; Wen, B.; Cheng, L. A novel polyaniline-coated bagasse fiber composite with core-shell heterostructure provides effective electromagnetic shielding performance. *ACS Appl. Mater. Interfaces* **2017**, *9*, 809–818. [CrossRef]
15. Wanasinghe, D.; Aslani, F.; Ma, G.; Habibi, D. Review of polymer composites with diverse nanofillers for electromagnetic interference shielding. *Nanomaterials* **2020**, *10*, 541. [CrossRef] [PubMed]
16. Kamchi, N.E.; Belaabed, B.; Wojkiewicz, J.L.; Lamouri, S.; Lasri, T. Hybrid polyaniline/nanomagnetic particles composites: High performance materials for EMI shielding. *J. Appl. Polym. Sci.* **2013**, *127*, 4426–4432. [CrossRef]
17. Zahari, M.H.B.; Guan, B.H.; Meng, C.E. Development and evaluation of $BaFe_{12}O_{19}$-PANI-MWCNT composite for electromagnetic interference (EMI) shielding. *Prog. Electromagn. Res. C* **2018**, *80*, 55–64. [CrossRef]
18. Niu, Y. Preparation of polyaniline/polyacrylate composites and their application for electromagnetic interference shielding. *Polym. Compos.* **2006**, *27*, 627–632. [CrossRef]
19. Niu, Y. Electromagnetic interference shielding with polyaniline nanofibers composite coatings. *Polym. Eng. Sci.* **2008**, *48*, 355–359. [CrossRef]
20. Shakir, M.F.; Rashid, I.A.; Tariq, A.; Nawab, Y.; Afzal, A.; Nabeel, M.; Naseem, A.; Hamid, U. EMI shielding characteristics of electrically conductive polymer blends of PS/PANI in microwave and IR region. *J. Electronic Mat.* **2020**, *49*, 1660–1665. [CrossRef]
21. Sobhaa, A.P.; Sreekalac, P.S.; Narayanankuttya, S.K. Electrical, thermal, mechanical and electromagnetic interference shielding properties of PANI/FMWCNT/TPU composites. *Prog. Org. Coat.* **2017**, *113*, 168–174. [CrossRef]
22. Raagulan, K.; Braveenth, R.; Kim, B.M.; Lim, K.J.; Lee, S.B.; Kim, M.; Chai, K.Y. An effective utilization of MXene and its effect on electromagnetic interference shielding: Flexible, free-standing and thermally conductive composite from MXene–PAT–poly(p-aminophenol)–polyaniline co-polymer. *RSC Adv.* **2020**, *10*, 1613–1633. [CrossRef]
23. Bhowmick, B.; Mollick, M.R.; Mondal, D.; Maity, D.; Bain, M.K.; Bera, N.K.; Rana, D.; Chattopadhyay, S.; Chakraborty, M.; Chattopadhyay, D. Poloxamer and gelatin gel guided polyaniline nanofibers: Synthesis and characterization. *Polym. Int.* **2014**, *63*, 1505–1512. [CrossRef]
24. Li, M.; Guo, Y.; Wei, Y.; MacDiarmid, A.G.; Lelkes, P.I. Electrospinning polyaniline-contained gelatin nanofibers for tissue engineering applications. *Biomaterials* **2006**, *27*, 2705–2715. [CrossRef] [PubMed]
25. Ostrovidov, S.; Ebrahimi, M.; Bae, H.; Nguyen, H.K.; Salehi, S.; Kim, S.B.; Kumatani, A.; Matsue, T.; Shi, X.; Nakajima, K.; et al. Gelatin-polyaniline composite nanofibers enhanced excitation-contraction coupling system maturation in myotubes. *ACS Appl. Mater. Interfaces* **2017**, *49*, 42444–42458. [CrossRef] [PubMed]
26. Xue, J.; Liu, Y.; Darabi, M.A.; Tu, G.; Huang, L.; Ying, L.; Xiao, B.; Wu, Y.; Xing, M.; Zhang, L.; et al. An injectable conductive Gelatin-PANI hydrogel system serves as a promising carrier to deliver BMSCs for Parkinson's disease treatment. *Mater. Sci. Eng. C* **2019**, *100*, 584–597. [CrossRef]
27. Blinova, N.V.; Trchová, M.; Stejskal, J. The polymerization of aniline at a solution–gelatin gel interface. *Eur. Polym. J.* **2009**, *45*, 668–673. [CrossRef]
28. Long, Y.; Huang, K.; Yuan, J.; Han, D.; Niu, L.; Chen, Z.; Gu, C.; Jin, A.; Duvail, J.L. Electrical conductivity of a single Au/polyaniline microfiber. *Appl. Phys. Lett.* **2006**, *88*, 162113. [CrossRef]

29. Gareev, K.G.; Bagrets, V.S.; Khmelnitskiy, I.K.; Ivanitsa, M.G.; Testov, D.O. Research and Development of "Gelatin–Conductive Polymer" Composites for Electromagnetic Compatibility. In Proceedings of the 2020 IEEE Conference of Russian Young Researchers in Electrical and Electronic Engineering (EIConRus), St. Petersburg/Moscow, Russia, 27–30 January 2020; pp. 1069–1072. [CrossRef]
30. Gareev, K.G.; Luchinin, V.V.; Sevost'yanov, E.N.; Testov, I.O.; Testov, O.A. Frequency dependence of an electromagnetic absorption coefficient in magnetic fluid. *Tech. Phys.* **2019**, *89*, 954–957. [CrossRef]
31. Haberko, J.; Raczkowska, J.; Bernasik, A.; Rysz, J.; Nocuń, M.; Nizioł, J.; Łużny, W.; Budkowski, A. Conductivity of thin polymer films containing polyaniline. *Mol. Cryst. Liq. Cryst.* **2008**, *485*, 48–55. [CrossRef]
32. Vanga Bouanga, C.; Fatyeyeva, K.; Baillif, P.-Y.; Bardeau, J.-F.; Khaokong, C.; Pilard, J.-F.; Tabellout, M. Study of dielectric relaxation phenomena and electrical properties of conductive polyaniline based composite films. *J. NonCrystal. Sol.* **2010**, *356*, 611–615. [CrossRef]
33. Altarawneh, M.M.; Alharazneh, G.A.; Al-Madanat, O.Y. Dielectric properties of single wall carbon nanotubes-based gelatin phantoms. *J. Adv. Dielectr.* **2018**, *8*, 1850010. [CrossRef]
34. Zhang, W.; Ye, H.-Y.; Graf, R.; Spiess, H.W.; Yao, Y.-F.; Zhu, R.-Q.; Xiong, R.-G. Tunable and Switchable Dielectric Constant in an Amphidynamic Crystal. *J. Am. Chem. Soc.* **2013**, *135*, 5230–5233. [CrossRef] [PubMed]
35. Ye, H.-Y.; Zhang, Y.; Fu, D.-W.; Xiong, R.-G. An above-room-temperature ferroelectric organo–metal halide perovskite: (3-Pyrrolinium)($CdCl_3$). *Angew. Chem. Int. Ed.* **2014**, *53*, 1–7. [CrossRef] [PubMed]

Article

Performance Study of Split Ferrite Cores Designed for EMI Suppression on Cables

Adrian Suarez [1,*], Jorge Victoria [1,2], Jose Torres [1], Pedro A. Martinez [1], Antonio Alcarria [2], Joaquin Perez [1], Raimundo Garcia-Olcina [1], Jesus Soret [1], Steffen Muetsch [2] and Alexander Gerfer [2]

1 Department of Electronic Engineering, University of Valencia, 46100 Burjassot, Spain; jorge.victoria@we-online.de (J.V.); jose.torres@uv.es (J.T.); pedro.a.martinez@uv.es (P.A.M.); joaquin.perez-soler@uv.es (J.P.); raimundo.garcia@uv.es (R.G.-O.); jesus.soret@uv.es (J.S.)

2 Würth Elektronik eiSos GmbH & Co. KG, 74638 Waldenburg, Germany; antonio.alcarria@we-online.de (A.A.); steffen.muetsch@we-online.de (S.M.); alexander.gerfr@we-online.de (A.G.)

* Correspondence: adrian.suarez@uv.es

Received: 14 October 2020; Accepted: 22 November 2020; Published: 24 November 2020

Abstract: The ideal procedure to start designing an electronic device is to consider the electromagnetic compatibility (EMC) from the beginning. Even so, EMC problems can appear afterward, especially when the designed system is interconnected with external devices. Thereby, electromagnetic interferences (EMIs) could be transmitted to our device from power cables that interconnect it with an external power source or are connected to another system to establish wired communication. The application of an EMI suppressor such as a sleeve core that encircles the cables is a widely used technique to attenuate EM disturbances. This contribution is focused on the characterization of a variation of this cable filtering solution based on openable core clamp or snap ferrites. This component is manufactured by two split parts pressed together by a snap-on mechanism which turns this into a quick, easy to install solution for reducing post-cable assembly EMI problems. The performance of three different materials, including two polycrystalline (MnZn and NiZn) materials and nanocrystalline (NC) solution, are analyzed in terms of effectiveness when the solid sleeve cores are split. The possibility of splitting an NC core implies an innovative technique due to the brittleness of this material. Thus, the results obtained from this research make it possible to evaluate this sample's effectiveness compared to the polycrystalline ones. This characterization is carried out by the introduction of different gaps between the different split-cores and analyzing their behavior in terms of relative permeability and impedance. The results obtained experimentally are corroborated with the results obtained by a finite element method (FEM) simulation model with the aim of determining the performance of each material when it is used as an openable core clamp.

Keywords: electromagnetic interference (EMI) suppressors; sleeve ferrite cores; cable filtering; nanocrystalline (NC); split-core; snap ferrite; gap; DC currents; relative permeability; impedance

1. Introduction

The control of EMI in electronic devices is an increasing issue faced by designers in order to ensure that devices comply with EMC requirements to operate simultaneously without inferring with each other. This fact is due to the trends towards higher component integration, printed circuit board (PCB) size and thickness reduction, and the miniaturization of the device housings. Besides, other factors, such as higher switching frequencies in power converters and communication data rates in digital circuits, could lead to EMI problems [1,2]. Consequently, EMC engineering should be handled with the system approach, considering EMC throughout the design process to prevent possible EMI problems that could degrade device performance. Therefore, adopting specific solutions as early as possible in the design stage to meet the EMC requirements is of primary importance to reduce penalties from the standpoint of costs, time-to-market, and performance [3,4]. The EMC testing process can reveal that the shielding of a certain cable is needed or may even detect an unexpected EMI source when the designed device is connected to external modules, such as a power supply or to another device to communicate with it. When the cables represent the EMI source, that implies failing the conducted or radiated emissions test, and a widely used technique is applying an EMI suppressor such as a cable ferrite [4].

A cable ferrite's effectiveness to reduce EMI in cables is defined by its capability to increase the flux density of a specific field strength created around a conductor. The presence of noise current in a conductor generates an undesired magnetic field around it that can result in EMI problems. When a cable ferrite is applied in the conductor, the magnetic field is concentrated into magnetic flux inside the cable ferrite because it provides a higher magnetic permeability than air. As a result, the flowing noise current in the conductor is reduced and, thus, the EMI is attenuated. Currents that flow in cables (with two or more conductors) can be divided into differential mode (DM) and common mode (CM) depending on the directions of propagation. Although DM currents are usually significantly higher than CM currents, one of the most common radiated EMI problems is originated by CM currents flowing through the cables of the system [5]. CM currents, despite not having a high value, have a much greater interfering potential. This fact is because only a few microamps are required to flow through a cable to fail radiated emission requirements [4]. The use of a cable ferrite is an efficient solution to filter the CM currents in cables because, if a pair of adjacent conductors is considered, when the cable ferrite is placed over both signal and ground wires, the CM noise is reduced. As shown in Figure 1, the CM currents in both wires flow in the same direction, so the two magnetic fluxes in the cable ferrite are added together, and the filtering action occurs. The intended (DM) current is not affected by the presence of the cable ferrite because the DM current travels in opposite directions and it is transmitted through the signal and returns. Thus, the currents of the two conductors are opposing, meaning they cancel out and the cable ferrite has no effect [6]. This ability to attenuate only the undesirable CM disturbances is a very interesting feature of this kind of component [7–9].

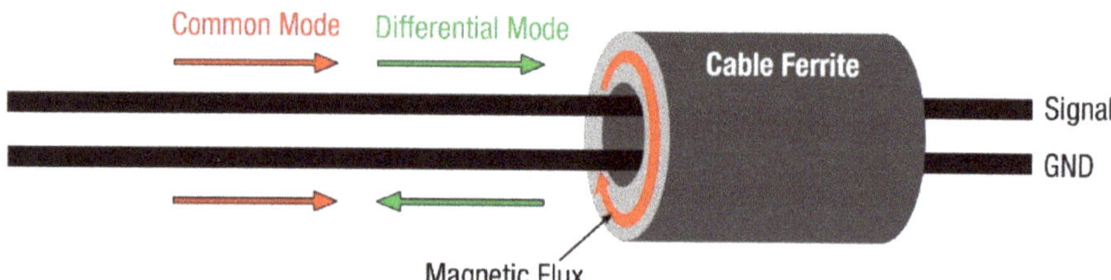

Figure 1. Diagram of common mode (CM) and differential mode (DM) currents passing through a cable ferrite with two adjacent conductors (signal and return paths).

This component represents a solution when the cables turn into an EMI source. It can be applied to peripheral and communication cables such as multiconductor USB or video cables to prevent interferences that could be propagated along the wire, affecting the devices that are interconnected [10,11]. This component is also widely used to reduce high-frequency oscillations caused by the increasingly fast switching in power inverters and converters with cables attached without scarifying the switching speed and increasing the power loss. Therefore, selecting the proper cable ferrite makes it possible to reduce the switching noise by increasing the propagation path impedance in the desired frequency range [5].

The application of cable ferrites is a widely used technique to reduce EMI in cables, despite the drawbacks that the integration of an extra component can involve in terms of cost and production of the system [12]. Nevertheless, these drawbacks are usually compensated by the effectiveness of cable ferrites to filter interferences without having to redesign the electronic circuit [6,7]. Manufacturers provide a wide range of ferrite cores with different shapes, dimensions and compositions, but the most widely used solutions applied to cables are the sleeve (or tube) ferrite cores or their split variation, the openable core clamp (or snap ferrites) [13]. This component is manufactured from two split parts pressed together by a snap-on mechanism, turning this into a quick, easy to install solution for reducing post-cable assembly EMI problems. The main advantage of snap ferrites compared to solid sleeve ferrite cores is the possibility to add them to the final design, without manufacturing a specific cable that includes the sleeve ferrite core before assembling its connectors.

Nevertheless, the halved ferrite's performance will be lower than that of a solid core with the same composition and dimensions in terms of the relative permeability (μr) and hence the impedance introduced in the cable [14]. This performance degradation is caused by the gap introduced between the split parts. Additionally, the presence of a defined air gap between the split parts can turn into an advantage from the standpoint of the core saturation because it allows for higher DC currents before saturation is reached as compared to solid cores. For applications such as power supplies or motor drivers, high DC currents flow through the cable, and the performance of the cable ferrite can be degraded [8]. Therefore, in these situations, it is interesting to halve the cable ferrite with the aim of introducing a controlled gap that reduces the influence of DC currents into it [15,16].

The materials selected to carry out the characterization in terms of cable ferrite performance considering gap and stability to DC currents are two ceramic cores based on MnZn and NiZn and a third core of nanocrystalline (NC) structure. One of the main advantages of MnZn and NiZn materials is the possibility of creating cores with many different shapes and the possibility of halving them without modifying their internal structure [17,18]. Preliminary studies have shown that NC sleeve ferrite cores provide a significant effectiveness when used as an EMI suppressor [17,19,20]. Nevertheless, the internal structure and manufacturing process have traditionally made it complicated to obtain a split-core sample that can be used as a snap ferrite, keeping its effectiveness. Therefore, a prototype of a split-core of an NC sample has been manufactured based on a new cutting and assembling technique that makes it possible to analyze this material's performance when it is halved.

Consequently, one of the main objectives of this contribution is to analyze the dependencies between the gap parameter and the performance in terms of impedance provided by the snap ferrite. This analysis is performed through an experimental measurement setup that is compared with the results obtained through a finite element method (FEM) simulation model. The simulation model helps to determine the study's accuracy, specifically in the high-frequency range where parasitic elements may affect the experimental results [20]. Likewise, the stability of three solid (not split) and split cable ferrites based on different compositions are characterized in order to determine the influence of DC currents on the impedance response provided. The results obtained from this study make it possible to compare the different materials to find out which one is the most efficient, depending on the frequency range.

Thereby, the three different materials and their structures are described in Section 2 through the main magnetic parameters, such as the relative permeability and the reluctance caused by introducing a gap. Section 3 defines the measurement setups employed to perform the experimental results and the designed FEM simulation model description. Subsequently, in Section 4, the three different samples' performance under test is shown in terms of impedance versus frequency. The dependencies on the air gap introduced in the split-cores and the influence on the injection of DC currents are discussed. Finally, the main conclusions are summarized in Section 5 to determine the performance of each material when used as an openable core clamp.

2. Magnetic Properties

The magnetically soft ferrites are widely used for manufacturing EMI suppressors as cable ferrites. Conventionally, the most used ferrite cores for filtering applications are based on ceramic materials (also known as polycrystalline materials) and, although they do not belong to the metals group, they are made from metal oxides such as ferrite, nickel and zinc. MnZn and NiZn represent two extensively used solutions due to their heat resistance, hardness, and high resistance to pressure. An advantage of ceramic materials is the possibility of manufacturing components with many different shapes and dimensions. The remarkable fact about the ceramic ferrites is that they combine extremely high electrical resistivity with reasonably good magnetic properties [21]. The starting material is iron oxide Fe_2O_3 that is mixed with one or more divalent transition metals, such as manganese, zinc, nickel, cobalt or magnesium [14]. The manufacturing procedure can be divided into these steps. First, the base materials are weighed into the desired proportions and wet mixed in ball mills to obtain a uniform distribution and particle size. Next, the water is removed in a filter press, and the ferrite is loosely pressed into blocks and dried. It is then pre-fired (calcined) at about 1000 °C to form the ferrite. The pre-sintered material is then milled to obtain a specific particle size. Subsequently, the dry powder is mixed with an organic binder and lubricants before being shaped by a pressing technique to obtain the final form. Finally, the resultant green core is subjected to a heating and cooling cycle, reaching temperatures higher than 1150 °C, promoting any unreacted oxides to be formed into ferrite. The manufacturing procedure and the material mix are essential to define a ceramic core's magnetic properties. With MnZn materials, it is possible to obtain samples that provide initial permeabilities of the order of 1000–20,000 and provide a low resistivity (0.1–100 Ω·m). Their range of frequency for EMI suppression applications covers from hundreds of kHz to some MHz.

Regarding NiZn materials, these provide initial permeabilities of the order of 100–2000, so they are intended for a higher frequency operation than MnZn, covering from tens of MHz up to several hundreds of MHz. In terms of resistivity, NiZn materials reach high values (about 10^4–10^6 Ω·m) [15,21,22]. Therefore, considering the structure of ceramic cores, they can be considered as isotropic.

The structure and manufacturing technique used for ceramics make it possible to produce split-cores or cut a solid core after its production with water-cooled diamond tools to build snap ferrites [14]. The NC cores' manufacturing procedure is quite different from the one used for ceramic production since they are formed by a continuous laminar structure that is wound to form the final core. The tape-wound structure is based on an amorphous ribbon of only 7–25 micrometers in thickness. It is generated by melting the base material by heating it at 1300 °C and depositing it on a high-speed cooling wheel (100 km/h) that reduces the temperature of the material to 20 °C at a rate of 10^6 K/s. After that, the rolled material is exposed to an annealing process, usually under a transversal and longitudinal magnetic field. This treatment affects the magnetic properties, resulting in ultrafine crystals with a size of the order of 7–20 nm. Finally, an epoxy coating or an additional protective case is needed to protect the sample, due to the brittle nature of the tape. Depending on the parameters selected during the manufacturing procedure, NC samples can provide initial permeability values in the range of 15,000 to 150,000. Electrical resistivity is relatively high even if it is considered a metallic material, generally over 10^{-6} Ω·m [14,15,17,23]. The NC material structure presents the advantage of designing smaller components with more significant magnetic properties for EMI suppression [19,20,24–27]. Figure 2 shows the NC core before adding the protective coating and its manufacturing procedure diagram.

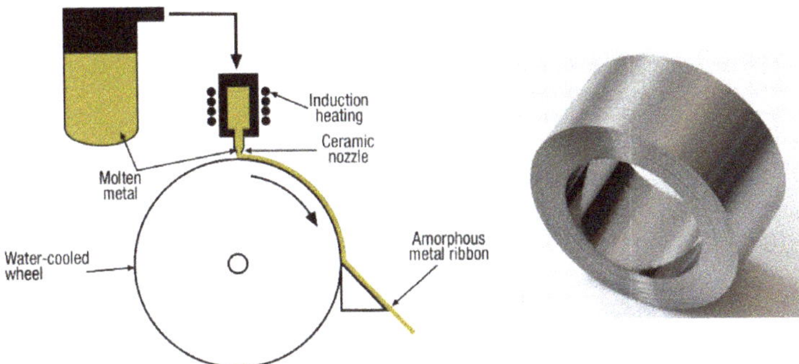

Figure 2. Nanocrystalline manufacturing procedure diagram and nanocrystalline (NC) final core sample without the protective coating.

Ceramic and NC samples can be manufactured as sleeve cores but have a different internal structure. From the point of view of the flux, it travels only in the rolling direction of the amorphous ribbon in the NC core. In ceramics, the flux is distributed uniformly because the material is a single homogeneous unit [21]. This different structure will result in a different behavior when the core is split into two parts for use as a snap ferrite. In ceramic cores, a performance reduction in terms of relative permeability when they are split is expected. However, we could anticipate that in the case of the NC structure, this decrease would be significant. This fact could be considered because when the NC sample is halved, the wound core is cut, limiting the flux path. Thereby, the halved faces have been plated in order to connect both halved parts with the aim of reducing the gap of the resultant snap core. Table 1 and Figure 3 show the dimensions of the samples used to develop characterization. Note that the split samples have the same dimensions as the solid cores.

Table 1. List of cable ferrite samples used in this research.

Magnetic Material	Outer Diameter (OD) (mm)	Inner Diameter (ID) (mm)	Height (H) (mm)	Thickness (mm)
NC	19.2	11.7	25.4	7.5
MnZn	18.6	10.2	25.2	8.4
NiZn	18.6	10.2	25.1	8.4

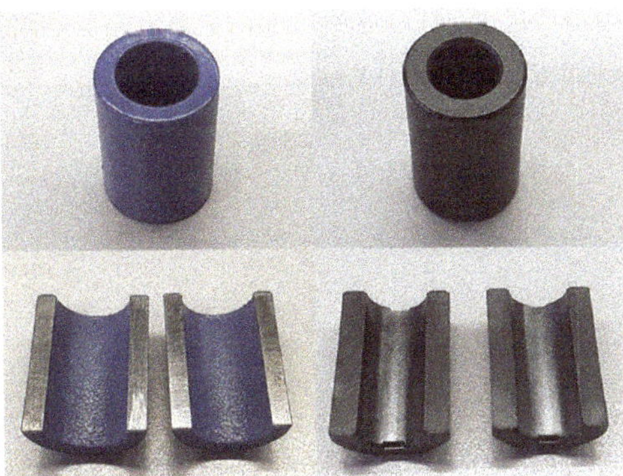

Figure 3. Non-split NC and split-core sample (**left**) and non-split ceramic and split-core sample (**right**).

The solid and split-cores' behavior based on the three different materials is analyzed in this section through the relative permeability. The permeability of magnetic materials generally depends on the magnetic flux density, DC bias currents, temperature, frequency, and intrinsic material properties [15,17]. When an air gap is included in a closed magnetic circuit, the circuit's total permeability is called the effective relative permeability μ_e and this is lower than the permeability of the original solid core without the presence of the air gap. In terms of EMI suppression, reducing the relative permeability in a cable ferrite is generally related to the decrease in its ability to attenuate interferences. However, the presence of an air gap is sometimes desired to increase the DC bias capability of the core or to reduce the permeability to achieve a more predictable and stable response with the aim of shifting the resonance frequency (f_r) to higher values to reduce the effects of dimensional effects [28,29].

When the two parts of the split-core are joined, a certain air gap remains between them that results in a magnetic reluctance (R) increase, since the gap represents an opposition to the magnetic flux (Φ) normal flow [15,30]. As shown in Figure 4, this effect is analogous to adding a series resistor in an electronic circuit to reduce the magnitude of the current. In Figure 4, R_c represents the reluctance of the core, R_g the reluctance of the gap, Φ the magnetic flux that flows through the magnetic path length of the core (l_c), l_g the length of the air gap, i the current that flows through the conductor and N the number of turns.

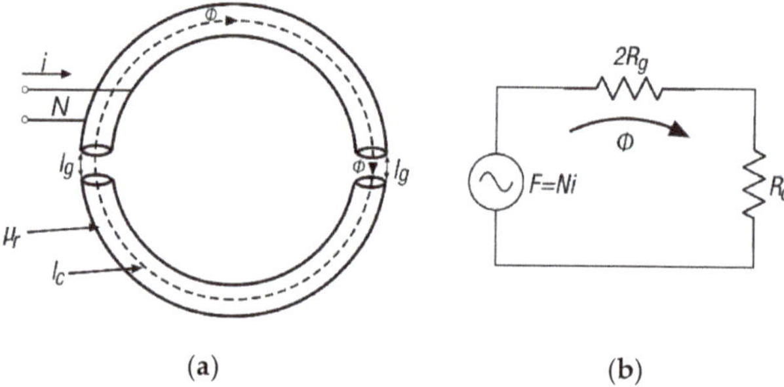

Figure 4. Split-core with air gaps: (**a**) Magnetic flux distribution diagram and (**b**) the magnetic circuit of a split-core with two air gaps.

The general expression to obtain the magnetic reluctance is given by [15]:

$$R = \frac{l}{\mu_r \mu_0 A} \left[H^{-1}\right] \quad (1)$$

where l corresponds to the magnetic path length and A to the cross-sectional area. The l and A parameters are obtained from the dimensional features of the sample, considering a toroid with a rectangular cross-section:

$$l = \pi\left(\frac{OD}{2} + \frac{ID}{2}\right)[m] \quad (2)$$

$$A = H\left(\frac{OD}{2} - \frac{ID}{2}\right)[m] \quad (3)$$

where H is the core's height and OD and ID are the outer and inner diameter, respectively. The overall reluctance of the split-core considering the air gap can be calculated from (1) as the sum of the reluctance core (R_c) and reluctance air gap (R_g) [13,15]:

$$R = R_c + 2R_g = \frac{l_c - l_g}{\mu_r \mu_0 A} + \frac{2l_g}{\mu_0 A} \left[H^{-1}\right] \quad (4)$$

thereby, the air gap factor (F_g) is

$$F_g = \frac{R}{R_c} = \frac{R_c + 2R_g}{R_c} = 1 + \frac{2R_g}{R_c} = 1 + \frac{\mu_r 2 l_g}{l_c} \tag{5}$$

and the effective relative permeability of a core with an air gap is [14,15,31]:

$$\mu_e = \frac{\mu_r}{1 + \frac{\mu_r 2 l_g}{l_c}} = \frac{\mu_r}{F_g}. \tag{6}$$

Equation (6) represents the most common and simplified model to approximate the effective permeability caused by an air gap since it underestimates the value of μ_e because it does not consider the effect of the fringing flux across the air gap [32]. In the case of toroid cores, to estimate the influence of the air gap introduced when the core is split into two parts, l_g is usually considered to be twice the spacer thickness [15,32]. In order to characterize the reduction of the relative permeability caused by an air gap, the three solid (not split) cores of Table 1 are compared with three split-cores of the same material and dimensions but introducing four different gaps. Thereby, five study cases are carried out for each core material:

1. Non-split-core: core without a gap.
2. Split-core $g0$: both parts of the core are joined without fixing a gap value. In order to differentiate this case with the non-split-core, a gap value of 0.01 mm is considered.
3. Split-core $g1$: both parts of the core are joined by fixing a gap value of 0.07 mm.
4. Split-core $g2$: both parts of the core are joined by fixing a gap value of 0.14 mm.
5. Split-core $g3$: both parts of the core are joined by fixing a gap value of 0.21 mm.

Figure 5 shows the experimental relative permeability measured for each of the three cores included in Table 1, considering the five different cases in terms of the gap introduced. The experimental traces (solid traces) are compared with the effective relative permeability calculated (dotted traces) by using Equation (6), considering the four gaps defined above. This parameter has been calculated from the experimental relative permeability of the non-split-core sample. These data are expressed through a vector formed by 801 frequency points with their corresponding permeability values. The effective relative permeability of a core with a specific gap is determined by computing these values point by point. Thereby, the air gap factor value F_g changes throughout the frequency range analyzed. It is possible to observe that both NiZn and MnZn graphs show a similar response between calculated and experimental results. There is a significant match in the low-frequency region, particularly in the NiZn samples, since they provide a lower and more stable permeability than MnZn cores. In the high-frequency region, the calculated effective relative permeability is lower than the experimentally measured one, verifying that Equation (6) provides an underestimation of this. Another difference between NiZn and MnZn traces is observed by comparing the $g0$ traces because the initial permeability decreases mostly in MnZn because the original non-split-core yields a higher initial permeability than the NiZn sample. The estimation of $g0$ traces was obtained by fixing a gap of 0.01 mm between both split parts in order to simulate the real snap ferrite's behavior, and the estimated values match with the experimental data [13]. The NC graph shows a different behavior than the ceramic core since the experimental and calculated permeability only matches in the low-frequency region. Therefore, unlike ceramic cores, it is not possible to estimate with Equation (6) the effective relative permeability of NC in the middle and high-frequency region when used as a snap ferrite due to its different internal structure.

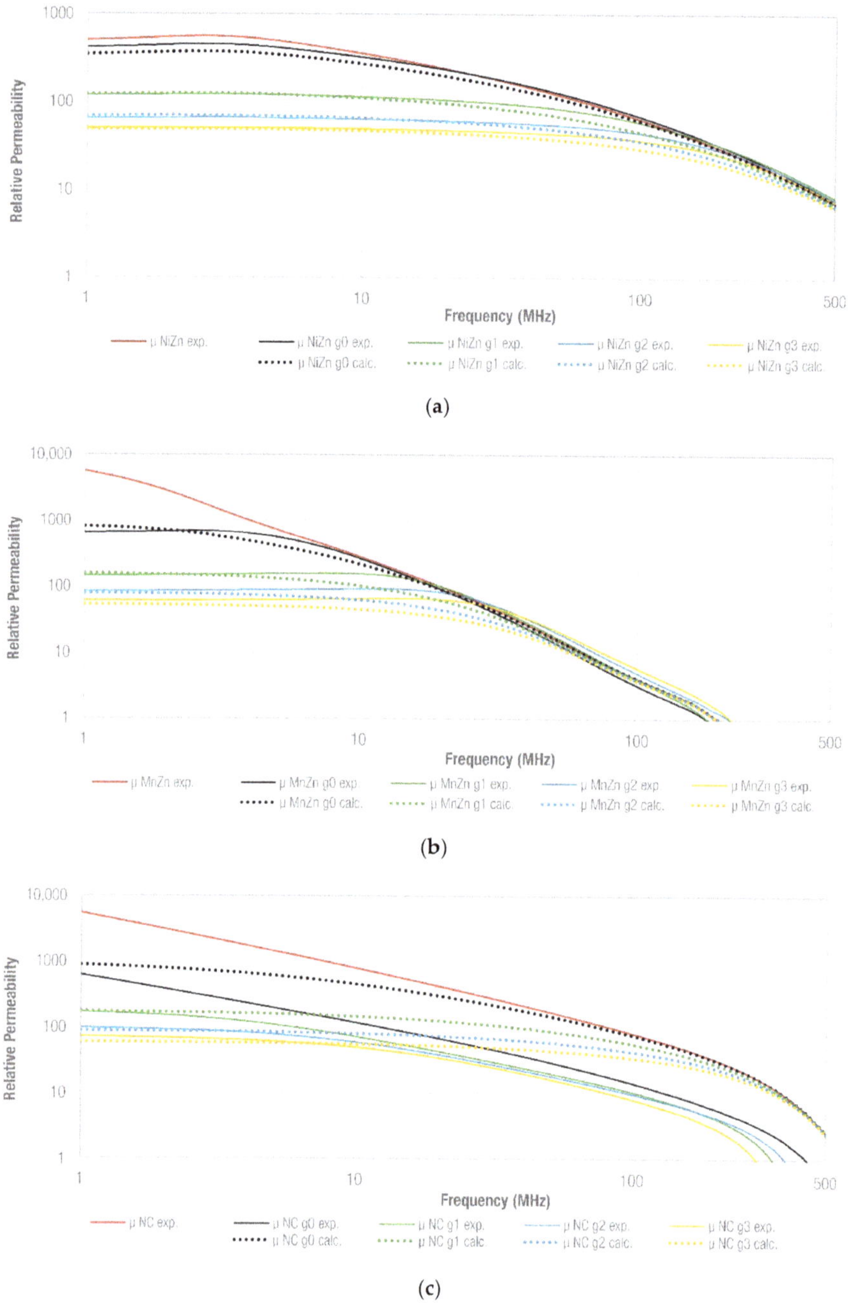

Figure 5. Comparison between experimental (solid traces) and estimated (dotted traces) effective permeability considering different gaps for NiZn, MnZn and NC samples: (**a**) NiZn non-split (red trace) and split cases; (**b**) MnZn non-split (red trace) and split cases and (**c**) NC non-split (red trace) and split cases.

3. Characterization Setups

EMI suppressors, such as cable ferrites, are usually classified by the impedance that they can introduce in a specific frequency range when they embrace a conductor. This parameter represents the magnitude of the impedance that can be represented from a series equivalent circuit mainly based on a resistive and inductive component [13,20]. The resistive component is connected to the imaginary part of the relative permeability representing the core's losses, whereas the real part of the permeability is related to the inductive component [22]. Therefore, there is a direct relationship between the core material's magnetic behavior and its performance in terms of impedance. Other factors that contribute to defining the impedance provided by a cable ferrite are the dimensions and the shape [19].

3.1. Impedance Measurement Setup

The experimental magnitude of the impedance of each sample is obtained by using the E4991A RF Impedance/Material Analyzer (Keysight, Santa Rosa, CA, USA) connected to the Terminal Adapter 16201A (Keysight, Santa Rosa, CA, USA). This adapter makes it possible to introduce into the measurement setup the 16200B External DC Bias (Keysight, Santa Rosa, CA, USA) that allows for supplying a bias current through the cable ferrite of up to 5 A using a 7 mm port and an external DC current source. Finally, the cable ferrite is connected by means of the Spring Clip Fixture 16092A (Keysight, Santa Rosa, CA, USA) that is connected to the 16200B test fixture [33]. After it is properly calibrated, this measurement setup is able to characterize cable ferrites from 1 MHz to 500 MHz since the E4991A equipment can operate from 1 MHz and the 16200B test fixture up to 500 MHz. Figure 6 shows the described experimental measurement setup.

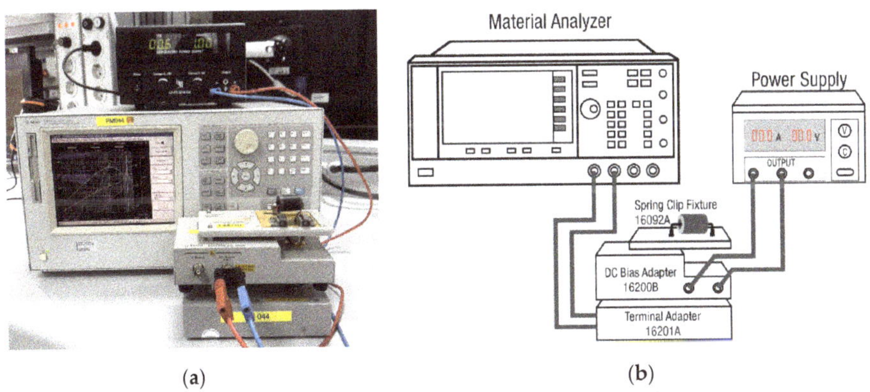

Figure 6. Impedance measurement setup with the DC bias test fixture connected: (**a**) Photograph of the measurement setup and (**b**) diagram blocks of the measurement setup.

This setup provides the experimental impedance of the split and non-split-cores, analyzing them when there is no presence of DC currents and increasing this parameter up to 5 A. The results obtained can be compared to analyze the behavior of each of the three materials characterized in this contribution in terms of the gap introduced in the core and the value of bias current injected.

3.2. Simulation Model

The different split and non-split cable ferrites' performance and the relationship between the impedance provided and the air gap introduced are specifically examined through an electromagnetic analysis simulator (Ansys Electronics Desktop). The proposed simulation model is shown in Figure 7. It is formed by a copper conductor that crosses a cylindrical core defined by the material properties of each of the materials described in Section 2. The conductor is connected to two ports (input and output) referenced to a perfect electrical plane located at a certain distance under it. This simulation setup represents a transmission line based on a parallel line (or single wire) considering a single wire over a ground plane that allows for designing a system with a characteristic impedance of $Z_0 = 150\ \Omega$. This parameter is fixed by selecting the distance from the plane to the center of the conductor $H = 15$ mm, the diameter of the conductor $d = 4.9$ mm and considering that it is surrounded by air [34–36]. By setting the ports' impedance to 150 Ω, it is possible to extract the cable ferrite's impedance under test without the characterization system influencing the results obtained. This value is a reference value adopted in different EMC standards to characterize and calibrate devices such as common mode absorption devices (CMADs) intended for measuring EMI disturbances in cables [11]. These fixtures are characterized using the through-reflect-line (TRL) calibration method based on measuring the S-parameters of CMADs, as described in CISPR 16-1-4 [11]. Therefore, this simulation model provides a reference system that can extract the impedance introduced in the conductor by the cable ferrite. The procedure performed to emulate the different studied gap cases ($g0$, $g1$, $g2$ and $g3$) is based on a parametric gap sweep. This technique makes it possible to determine the sleeve core's impedance when it is split into two parts and a specific gap is introduced. It is expected that this simulation model is able to provide the performance of the split samples from the original relative permeability (the values obtained for the non-split-core sample) by fixing the gaps described in Section 2. In the $g0$ case, the 0.01 mm distance value was introduced to differentiate it from the original non-split core.

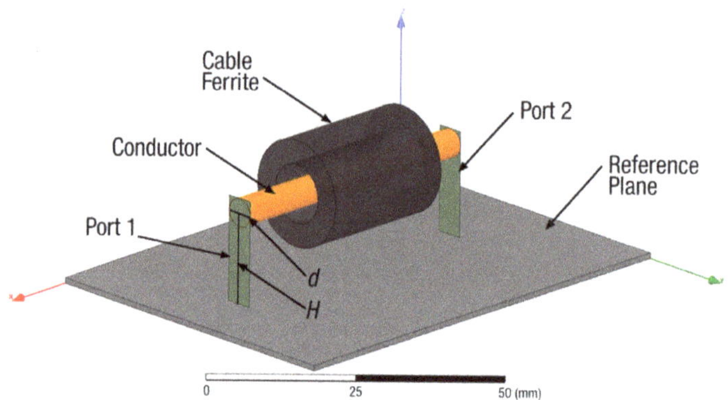

Figure 7. Cable ferrite simulation model.

4. Results and Discussion

This section focuses on analyzing the EMI suppression performance of the three described materials when they are split to be employed as snap ferrites. The results obtained from the experimental measurement setup and those provided by the simulation model are compared through each materials' impedance. This comparison is carried out to verify that the experimental results are not influenced by elements such as stability of the calibration setup in the high-frequency region and undesired high-frequency resonances caused by parasitic elements that could reduce the accuracy of the measurement. As is shown in Figure 8, the results are organized in three graphs, one per material: NiZn (a), MnZn (b) and NC (c)). These graphs represent the experimental (solid traces) and computed (dotted traces) impedance provided by the cable ferrite, considering the non-split situation and the split cases where the core is separated into two parts.

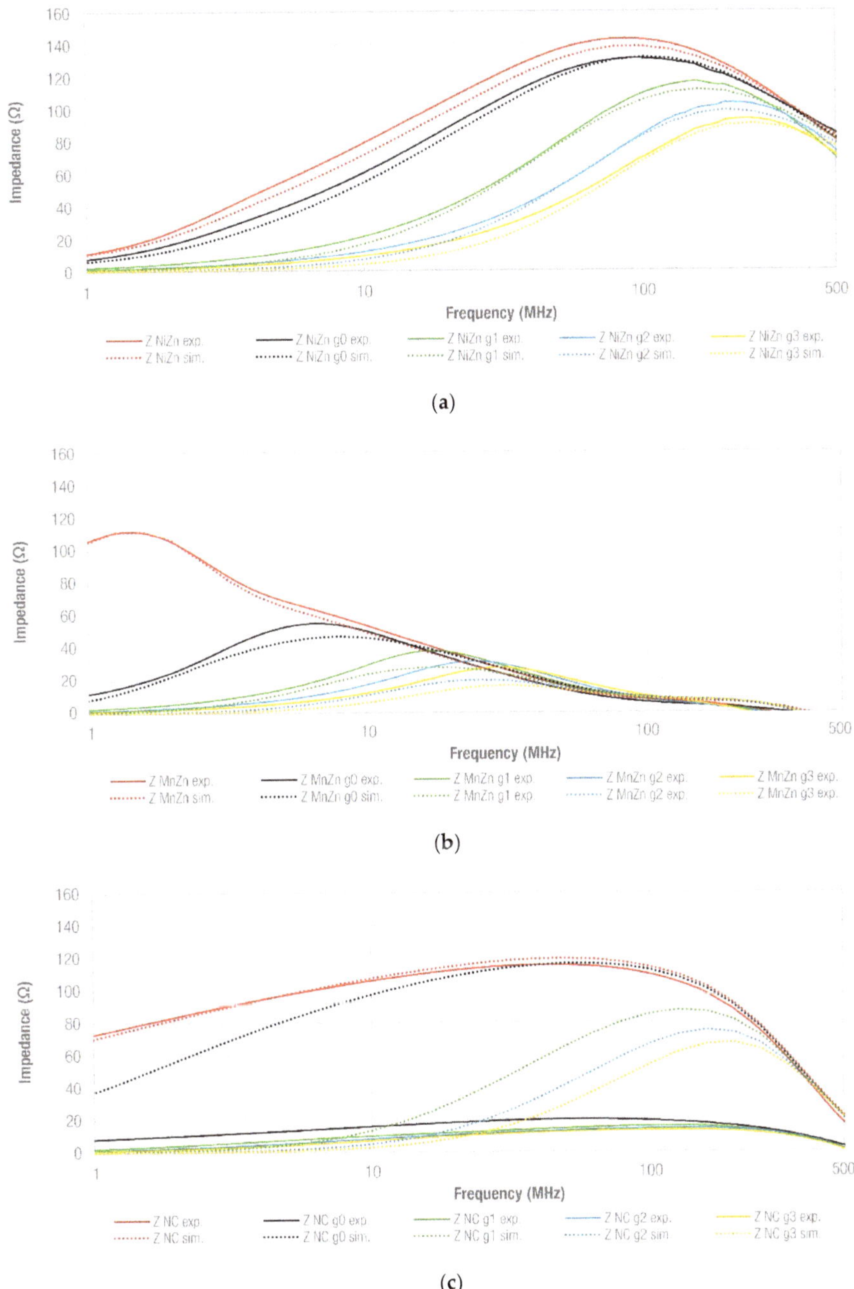

Figure 8. Comparison between experimental (solid traces) and simulated (dotted traces) impedance considering different gaps (non-split and split with gaps $g0$, $g1$, $g2$ and $g3$) for NiZn, MnZn and NC cable ferrites: (**a**) NiZn non-split (red trace) and split cases; (**b**) MnZn non-split (red trace) and split cases and (**c**) NC non-split (red trace) and split cases.

From the results obtained, the red traces of the graphs show the impedance provided by each non-split sample and it is possible to verify that the simulated and experimental results are a good match and, consequently, the data derived from the experimental setup can be considered in the whole frequency range analyzed. Consequently, a parametric gap sweep was performed in the simulation model by setting the four defined gap situations ($g0$, $g1$, $g2$ and $g3$) and keeping the same magnetic properties introduced to the non-split model. As can be observed, there is an excellent agreement between simulated and experimentally obtained results in NiZn and MnZn traces, whereas there is a significant difference in the NC case. This fact correlates with the conclusions obtained from the effective permeability data of the NC sample, shown in Section 2. Therefore, it is not possible to determine the NC sample's behavior when it is split and gapped by considering the non-split sample's magnetic properties. This is because the cut section's metallization is not able to maintain the high performance of the original NC sample. Then, the NC experimental results are considered to compare its performance with that provided by NiZn and MnZn samples.

Based on the results of three materials, halving the cable ferrite and using it as a snap ferrite ($g0$ situation) results in a shift of the resonance frequency at which the sample is able to provide the maximum attenuation ratio at the same time that the impedance is reduced. From the standpoint of the equivalent inductance and resistance series circuit of the sleeve core, the f_r is produced when the inductive component (X_L) turns into negative values and the resistive part (R) reaches the maximum value. Above this frequency value, the sleeve core's performance is degraded by the parasitic capacitive effect. Therefore, the f_r to higher frequencies shift results in extending the frequency range in which the sleeve core is effective to reduce EMI. This effect is lower for the NiZn cable ferrite since the f_r is increased from 86.7 MHz to 92.3 MHz, providing 142.9 Ω and 130.7 Ω, respectively. It represents a reduction of 8.5% in terms of impedance and an increase in the resonance frequency of f_r = 5.6%. Regarding the results obtained when a gap is introduced, an impedance of Z = 116.3 Ω (34.9% reduction) at f_r = 152.2 MHz for the $g1$ case, Z = 103.3 Ω (27.7% reduction) at f_r = 199.0 MHz for the $g2$ case and Z = 93.0 Ω (27.7% reduction) at f_r = 240.6 MHz for the $g3$ case. In the case of MnZn, it is possible to observe that the impedance traces are significantly modified when the sample is split since the original sample provides a maximum value of Z = 111.7 Ω at f_r = 1.4 MHz. The split-core with one part attached to the other ($g0$ case) provides Z = 55.0 Ω at f_r = 6.6 MHz, so the performance of the cable ferrite is reduced by about 50.8%. It is a relevant performance reduction compared to the attenuation ratio reduction of the NiZn sample. For the rest of the MnZn study cases where a higher gap is introduced ($g1$, $g2$ and $g3$), the impedance is mostly reduced (66.1%, 58.0% and 75.6%, respectively), compared to the NiZn results. Regarding the NC results, as described above, the simulated results obtained for the non-split sample match significantly with the experimental ones. Nevertheless, when the core is split, the simulated results overestimate the experimental data since the maximum impedance provided by the not split sample corresponds to 115.0 Ω, whose f_r = 45.5 MHz, whereas when it is split with both parts attached as closely as possible, the impedance is reduced by about 82.9%. When a specific gap is introduced ($g1$, $g2$ and $g3$ cases), the attenuation ratio produced is 86–89%.

The NC simulation model magnetic parameters were modified with the objective of obtaining a more realistic approximation response. Thereby, the model was simulated considering three different situations: non-split-core for the original sample, split-core without introducing a gap ($g0$) and split-core with the intended gap ($g1$, $g2$ and $g3$). Consequently, the magnetic parameters of the $g0$ situation correspond to the effective permeability measured with the split-core with both parts attached. The rest of gapped cases ($g1$, $g2$ and $g3$) were simulated by considering the measured effective permeability of the sample when the gap $g1$ is introduced. Figure 9 shows that the new simulated results match significantly with the experimental traces.

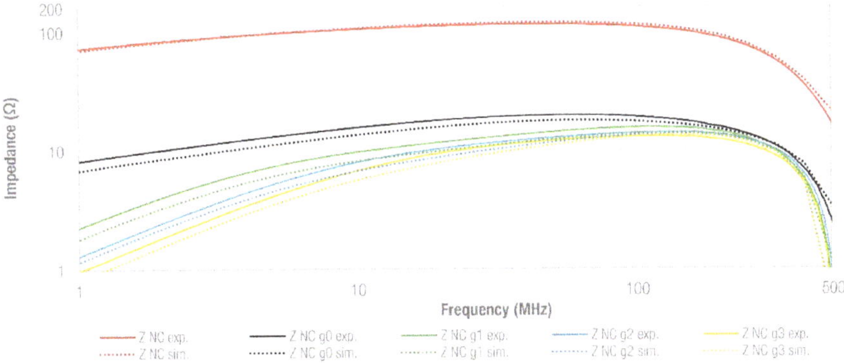

Figure 9. Comparison between experimental (solid traces) and new simulated (dotted traces) impedance, considering different gaps (non-split and split with gaps $g0$, $g1$, $g2$ and $g3$) for NC cable ferrites.

The comparison of the three different cable ferrites by separating them depending on the gap introduced is shown in Figure 10. As can be observed in Figure 10a, when cores are not split, they can be divided into three frequency ranges based on their performance. As expected, MnZn provides the larger impedance value in the low-frequency region, yielding the best performance up to 2.9 MHz. The NiZn cable ferrite offers higher impedance than MnZn and NC samples above 23.4 MHz, representing the most effective solution to reduce EMI disturbances in the high-frequency region. The NC core offers excellent performance in the medium-frequency region, providing a great impedance throughout the frequency band from 2.9 MHz to 23.4 MHz.

Additionally, the non-split NC core is able to yield a more stable response up to its maximum impedance value and it shows a better performance than ceramic cores to reduce EMI disturbances when they are distributed in a wideband frequency range (from the low-frequency region up to about 100 MHz). Figure 10b shows the impedance comparison when the cores are split into two parts and attached as closely as possible, emulating a snap ferrite's function. In this case, NC has significantly reduced its performance and MnZn provides the best performance up to 8.4 MHz. From this frequency value, NiZn yields the highest impedance value. In the rest of the analyzed gaps ($g1$, $g2$ and $g3$), the NiZn sample mainly represents the most interesting solution because MnZn and NC cable ferrites offer a lower impedance response. Thus, when a significant gap is introduced, the material with lower permeability is able to yield the best EMI attenuation.

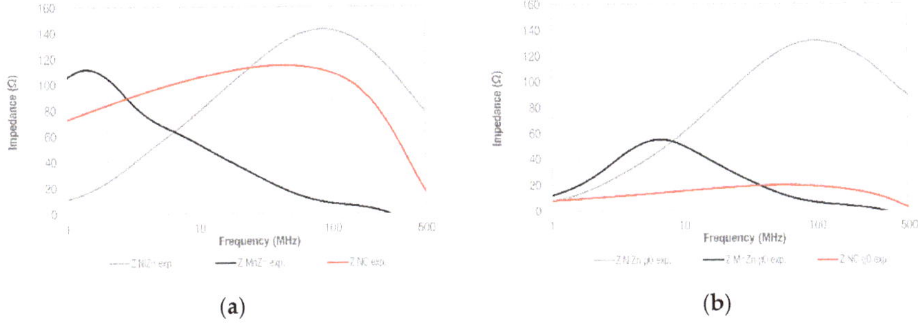

Figure 10. *Cont.*

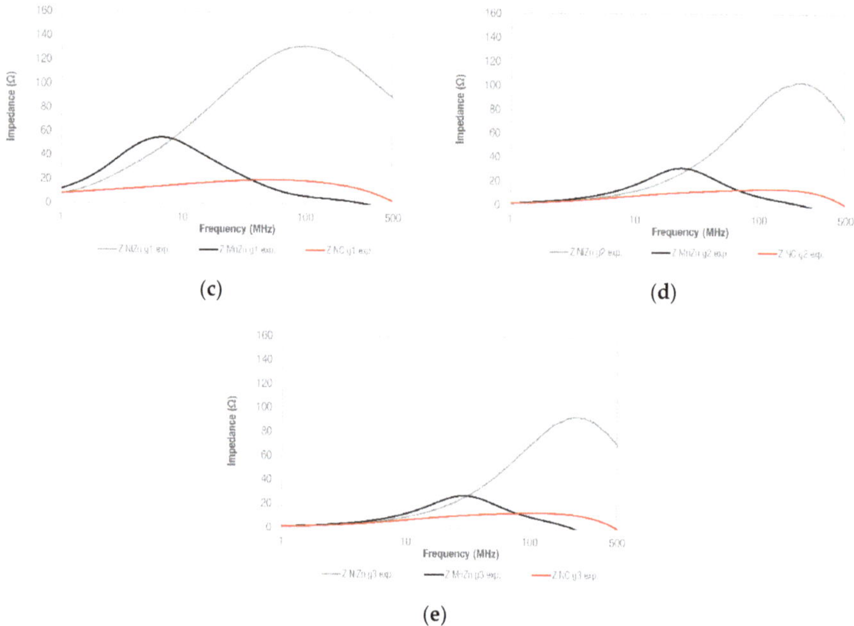

Figure 10. Comparison between the measured impedance of NiZn, MnZn and NC cable ferrites, considering five different gap cases: (**a**) NiZn, MnZn and NC non-split-cores; (**b**) NiZn, MnZn and NC $g0$ split-cores; (**c**) NiZn, MnZn and NC $g1$ split-cores; (**d**) NiZn, MnZn and NC $g2$ split-cores and (**e**) NiZn, MnZn and NC $g3$ split-cores.

Additionally, the non-split NC core is able to yield a more stable response up to its maximum impedance value and it shows a better performance than ceramic cores to reduce EMI disturbances when they are distributed in a wideband frequency range (from the low-frequency region up to about 100 MHz). Figure 10b shows the impedance comparison when the cores are split into two parts and attached as closely as possible, emulating a snap ferrite's function. In this case, NC has significantly reduced its performance and MnZn provides the best performance up to 8.4 MHz. From this frequency value, NiZn yields the highest impedance value. In the rest of the analyzed gaps ($g1$, $g2$ and $g3$), the NiZn sample mainly represents the most interesting solution because MnZn and NC cable ferrites offer a lower impedance response. Thus, when a significant gap is introduced, the material with lower permeability is able to yield the best EMI attenuation.

How splitting a cable ferrite into two parts, to be employed as a snap ferrite, modifies the impedance behavior was analyzed. Depending on the core's magnetic properties and structure, this involves a certain degradation of the EMI suppression ability. Nevertheless, splitting a cable ferrite could result in an advantage if the component is intended to encircle cables where DC currents are flowing. To further investigate this effect of the DC currents on the impedance, Figure 11 shows the impedance response of each of the three different materials studied when they are under DC bias conditions. Each material is represented in a separate graph and the response of the non-split sample is shown when different values of DC currents are injected (0 A, 0.5 A, 1 A, 2 A and 5 A), as described in Section 3. Figure 11a shows that the five traces have a similar trend, but the higher the DC current value, the lower the sample's impedance. This effect is observed from the lowest DC current value (0.5 A) and does not modify the impedance response significantly when compared to MnZn and NC results (Figure 11b,c, respectively). It is interesting how the resonance frequency is moved to higher frequencies when an increasing DC current is injected into the MnZn sample, specifically in the cases of 2 A and 5 A.

As regards NC behavior, it shows a significant impedance decrease in the low-frequency region and its performance is reduced more than that of ceramic cores when a DC bias flows through the cable.

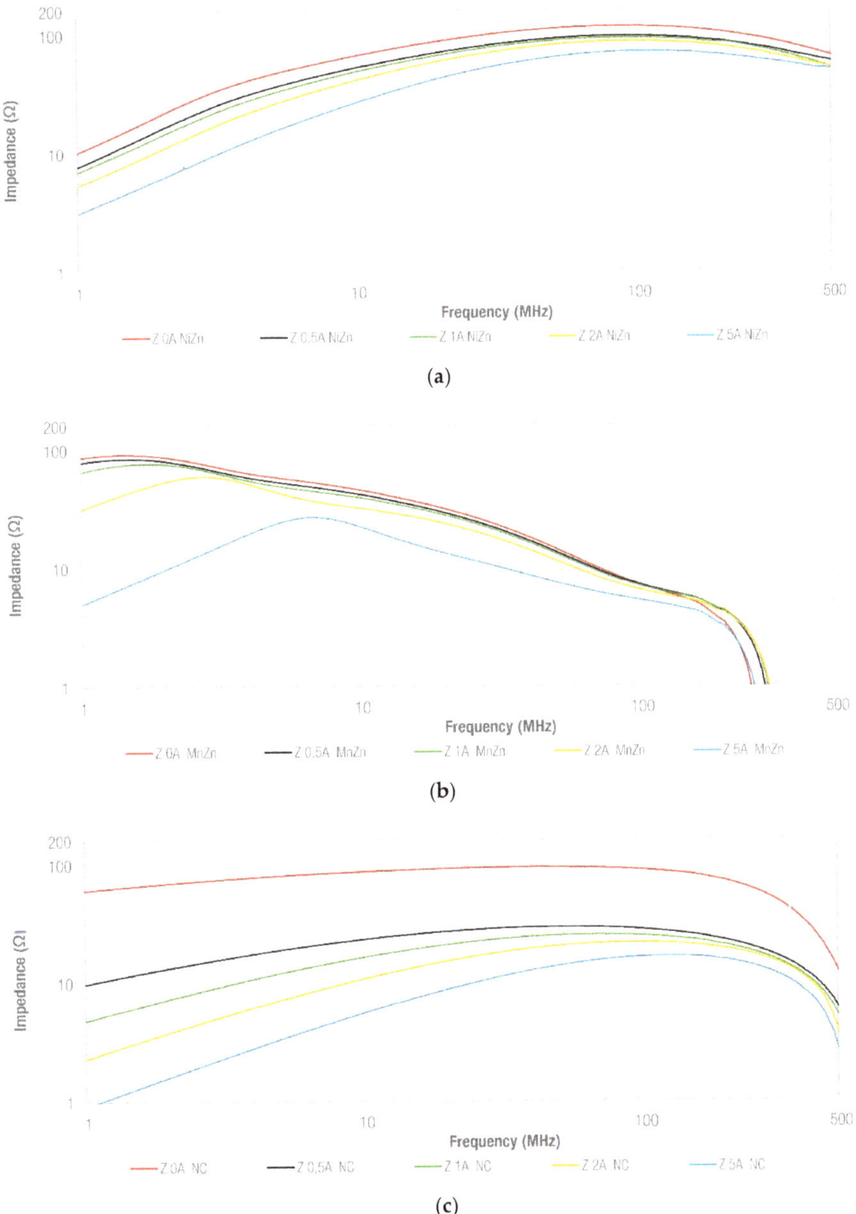

Figure 11. Impedance analysis considering different values of DC currents for the three different non-split samples: (**a**) NiZn; (**b**) MnZn and (**c**) NC.

The same analysis is repeated in Figure 12 for different gap values introduced in each of the samples. Thereby, the first row corresponds to the split NiZn samples, the second to the split MnZn samples and the third to the split NC samples. The left column shows the behavior of each material when the *g*0 gap is considered, and the right column shows the results obtained when *g*1 (dotted traces), *g*2 (short dashed traces) and *g*3 (long dashed traces) gaps are introduced. When the effect on the impedance response of splitting the samples to be attached without an intended gap is observed, the *g*0 traces show quite similar behavior in the three materials. NC traces have the same behavior over most of the frequency range, whereas NiZn traces show the same match between traces except for the 5 A case. In the case of MnZn, there is a difference between traces in the low-frequency region, producing a shift of the resonance frequency when DC currents higher than 0.5 A are applied. When the rest of the gaps are analyzed, the three materials have the same response when DC currents up to 5 A are injected. Moreover, from a certain frequency value, the materials' traces match independently of the DC current value and gap introduced.

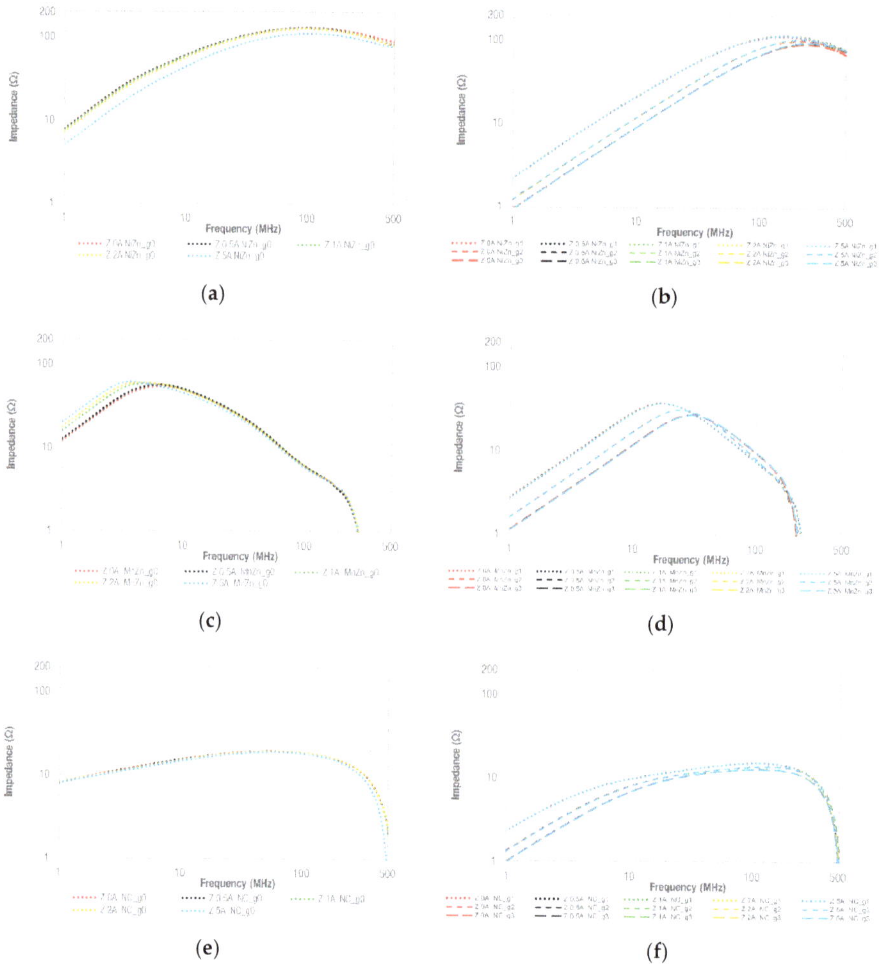

Figure 12. Impedance analysis considering different values of DC currents and gap conditions for the split samples: (**a**) NiZn *g*0 case; (**b**) NiZn *g*1, *g*2 and *g*3 cases; (**c**) MnZn *g*0 case; (**d**) MnZn *g*1, *g*2 and *g*3 cases; (**e**) NC *g*0 case and (**f**) NC *g*1, *g*2 and *g*3 cases.

5. Conclusions

The performance of the three different materials to build up ferrite cores was evaluated when they are split in order to determine their EMI suppression ability to be used as an openable core clamp. When the samples are not split, the analysis carried out in terms of impedance provided by each sample reveals that a ferrite core based on MnZn yields the best performance in the low-frequency region, whereas an NC core is most effective in the medium-frequency range and the NiZn sample provides larger impedance values in the high-frequency region.

When the samples are split and attached without introducing any gap ($g0$ situation), the impedance yielded by the NiZn sample is less degraded than the MnZn and NC impedances. In this study case, MnZn provides the best behavior in the low- and medium-frequency range, whereas the NC sample offers lower performance than expected due to its different internal structure. When larger gaps are considered, NiZn shows the most effective solution in terms of impedance. In this framework, other manufacturing procedures for NC snap-on cores should be investigated to obtain similar performance to what this solution can offer when it is not split.

The results obtained from the transmission line simulation model verify that the experimental results are in agreement and, thus, the data derived from the experimental measurement setup can be considered as an accurate approach in the frequency range studied (1–500 MHz). Consequently, the experimental and simulated results coincide with the conclusions obtained from effective relative permeability data. The material with more stable properties can provide higher performance and more predictable behavior than those with greater magnetic properties when the core is split. This conclusion is also applied when DC currents are flowing through the cable to be shielded since the NiZn solution shows better stability than the other materials.

Author Contributions: Conceptualization, A.S., J.V. and J.T.; Formal analysis, A.S., P.A.M., R.G.-O. and J.S.; Investigation, A.S., J.V., J.T., P.A.M. and A.A.; Methodology, A.S., P.A.M. and A.A.; Project administration, J.V., J.T., J.S., S.M. and A.G.; Supervision, J.V., J.T., J.S. and S.M.; Writing—original draft, A.S., J.T., P.A.M., J.P. and R.G.-O.; Writing—review and editing, J.V., J.T., P.A.M., A.A., J.P., R.G.-O., J.S., S.M. and A.G. All authors have read and agreed to the published version of the manuscript.

Funding: The APC was funded by the Universitat de València.

Acknowledgments: This work was supported by the Catedra Würth-EMC, a research collaboration agreement between the University of Valencia and Würth Elektronik eiSos GmbH & Co. KG.

Conflicts of Interest: The authors declare no conflict of interest. The founding sponsors had no role in the design of the study; in the collection, analyses, or interpretation of data; in the writing of the manuscript, and in the decision to publish the results.

References

1. González-Vizuete, P.; Domínguez-Palacios, C.; Bernal-Méndez, J.; Martín-Prats, M.A. Simple Setup for Measuring the Response to Differential Mode Noise of Common Mode Chokes. *Electronics* **2020**, *9*, 381. [CrossRef]
2. Kim, J.; Rotaru, M.D.; Baek, S.; Park, J.; Iyer, M.K.; Kim, J. Analysis of noise coupling from a power distribution network to signal traces in high-speed multilayer printed circuit boards. *IEEE Trans. Electromagn. Compat.* **2006**, *48*, 319–330. [CrossRef]
3. Crovetti, P.S.; Musolino, F. Interference of Spread-Spectrum EMI and Digital Data Links under Narrowband Resonant Coupling. *Electronics* **2020**, *9*, 60. [CrossRef]
4. Ott, H.W. *Electromagnetic Compatibility Engineering*; John Wiley & Sons: Hoboken, NJ, USA, 2009.
5. Yao, J.; Li, Y.; Zhao, H.; Wang, S. Design of CM Inductor Based on Core Loss for Radiated EMI Reduction in Power Converters. In Proceedings of the 2019 IEEE Applied Power Electronics Conference and Exposition (APEC), Anaheim, CA, USA, 17–21 March 2019; pp. 2673–2680. [CrossRef]
6. Kraftmakher, Y. Experiments on ferrimagnetism. *Eur. J. Phys.* **2012**, *34*, 213. [CrossRef]
7. Williams, T. *EMC for Product Designers*; Elsevier Science & Technology: Burlington, MA, USA, 2006; pp. 361–364. [CrossRef]
8. Goldman, A. *Modern Ferrite Technology*, 2nd ed.; Springer: Pittsburgh, PA, USA, 2006.

9. Bondarenko, N.; Shao, P.; Orlando, A.; Koledintseva, M.Y.; Beetner, D.G.; Berger, P. Prediction of common-mode current reduction using ferrites in systems with cable harnesses. In Proceedings of the 2012 IEEE International Symposium on Electromagnetic Compatibility, Pittsburgh, PA, USA, 6–10 August 2012; pp. 80–84. [CrossRef]
10. Paul, C.R. *Introduction to Electromagnetic Compatibility*, 2nd ed.; Wiley Interscience: Hoboken, NJ, USA, 2006.
11. Urabe, J.; Fujii, K.; Dowaki, Y.; Jito, Y.; Matsumoto, Y.; Sugiura, A. A Method for Measuring the Characteristics of an EMI Suppression Ferrite Core. *IEEE Trans. Electromagn. Compat.* **2006**, *48*, 774–780. [CrossRef]
12. Tong, X.C. *Advanced Materials for Electromagnetic Interference Shielding*, 1st ed.; CRC Press: Boca Raton, FL, USA, 2009.
13. Orlando, A.; Koledintseva, M.Y.; Beetner, D.G.; Shao, P.; Berger, P. A lumped-element circuit model of ferrite chokes. In Proceedings of the 2010 IEEE International Symposium on Electromagnetic Compatibility, Fort Lauderdale, FL, USA, 25–30 July 2010; pp. 754–759. [CrossRef]
14. Van den Bossche, A.; Valchev, V.C. *Inductors and Transformers for Power Electronics*; CRC Press: Boca Raton, FL, USA, 2005.
15. Kazimierczuk, M.K. *High-Frequency Magnetic Components*; John Wiley & Sons: Hoboken, NJ, USA, 2009.
16. Weinschrott, A. New measurement method for high frequency cable mounted ferrites. In Proceedings of the 2005 International Symposium on Electromagnetic Compatibility, Chicago, IL, USA, 8–12 August 2005; pp. 312–314. [CrossRef]
17. Suarez, A.; Victoria, J.; Alcarria, A.; Torres, J.; Martinez, P.A.; Martos, J.; Soret, J.; Garcia-Olcina, R.; Muetsch, S. Characterization of Different Cable Ferrite Materials to Reduce the Electromagnetic Noise in the 2–150 kHz Frequency Range. *Materials* **2018**, *11*, 174. [CrossRef] [PubMed]
18. Damnjanovi, M.; Stojanovi, G.; Živanov, L.; Desnica, V. Comparison of different structures of ferrite EMI suppressors. *Microelectron. Int.* **2006**, *23*, 42–48. [CrossRef]
19. Suarez, A.; Victoria, J.; Martinez, P.A.; Alcarria, A.; Torres, J.; Molina, I. Analysis of different Sleeve Ferrite Cores Performance according to their Dimensions. In Proceedings of the 2019 International Symposium on Electromagnetic Compatibility-EMC EUROPE, Barcelona, Spain, 2–6 September 2019; pp. 88–93. [CrossRef]
20. Suarez, A.; Victoria, J.; Torres, J.; Martinez, P.A.; Alcarria, A.; Martos, J.; Garcia-Olcina, R.; Soret, J.; Muetsch, S.; Gerfer, A. Effectiveness Assessment of a Nanocrystalline Sleeve Ferrite Core Compared with Ceramic Cores for Reducing Conducted EMI. *Electronics* **2019**, *8*, 800. [CrossRef]
21. Cullity, B.D.; Graham, C.D. *Introduction to Magnetic Materials*; Wiley: Hoboken, NJ, USA, 2009.
22. Kącki, M.; Rylko, M.S.; Hayes, J.G.; Sullivan, C.R. Magnetic material selection for EMI filters. In Proceedings of the 2017 IEEE Energy Conversion Congress and Exposition (ECCE), Cincinnati, OH, USA, 1–5 October 2017; pp. 2350–2356. [CrossRef]
23. Cuellar, C. HF Characterization and Modeling of Magnetic Materials for the Passive Components Used in EMI Filters. Ph.D. Thesis, University of Lille, Lille, France, 2013.
24. Sixdenier, F.; Morand, J.; Salvado, A.; Bergogne, D. Statistical study of nanocrystalline alloy cut cores from two different manufacturers. *IEEE Trans. Magn.* **2014**, *50*, 1–4. [CrossRef]
25. Brander, T.; Gerfer, A.; Rall, B.; Zenkner, H. *Trilogy of Magnetics: Design Guide for EMI Filter Design, SMP & RF Circuits*, 4th ed.; Swiridoff Verlag: Künzelsau, Germany, 2010.
26. Roc'h, A.; Leferink, F. Nanocrystalline core material for high-performance common mode inductors. *IEEE Trans. Electromagn. Compat.* **2012**, *54*, 785–791. [CrossRef]
27. Thierry, W.; Thierry, S.; Benoît, V.; Dominique, G. Strong volume reduction of common mode choke for RFI filters with the help of nanocrystalline cores design and experiments. *J. Magn. Magn. Mater.* **2006**, *304*, 847–849. [CrossRef]
28. Brockman, F.G.; Dowling, P.H.; Steneck, W.G. Dimensional effects resulting from a high dielectric constant found in a ferromagnetic ferrite. *Phys. Rev.* **1950**, *77*, 85–93. [CrossRef]
29. Ponomarenko, N. Study of Frequency and Microstructure Dependencies of Magnetic Losses of Ferrite Materials and Components. Ph.D. Thesis, Riga Technical University, Riga, Latvia, 2014.
30. Fiorillo, F. *Measurement and Characterization of Magnetic Materials*; Elsevier: Amsterdam, The Netherlands, 2004.
31. Bhuiyan, R.H.; Dougal, R.A.; Ali, M. A miniature energy harvesting device for wireless sensors in electric power system. *IEEE Sens. J.* **2010**, *10*, 1249–1258. [CrossRef]
32. McLyman, C.W.T. *Transformer and Inductor Design Handbook*; Dekker: New York, NY, USA, 1988.

33. Keysight 16200B External DC Bias Adapter Operation and Service Manual. Available online: http://literature.cdn.keysight.com/litweb/pdf/16200B.pdf (accessed on 28 September 2020).
34. Chang, K. *Handbook of Microwave and Optical Components, Fiber and Electro-Optical Components*; Wiley: Hoboken, NJ, USA, 1991.
35. International Telephone, & Telegraph Corporation. *Reference Data for Radio Engineers*; Howard W. Sams and Company: Indianapolis, IN, USA, 1968.
36. Haase, H.; Nitsch, J.; Steinmetz, T. Transmission-line super theory: A new approach to an effective calculation of electromagnetic interactions. *URSI Radio Sci. Bull.* **2003**, *307*, 33–60. [CrossRef]

Article

Simple Setup for Measuring the Response to Differential Mode Noise of Common Mode Chokes

Pablo González-Vizuete [1], Carlos Domínguez-Palacios [1], Joaquín Bernal-Méndez [2,*] and María A. Martín-Prats [1]

1 Dpto.Ingeniería Electrónica, Escuela Técnica Superior de Ingeniería, Universidad de Sevilla, 41092 Seville, Spain; pgonzalez17@us.es (P.G.-V.); cardompal@alum.us.es (C.D.-P.); mmprats@us.es (M.A.M.-P.)

2 Dpto.Física Aplicada III, Escuela Técnica Superior de Ingeniería, Universidad de Sevilla, 41092 Seville, Spain

* Correspondence: jbmendez@us.es; Tel.: +34-954-486191

Received: 15 January 2020; Accepted: 21 February 2020; Published: 25 February 2020

Abstract: This work presents a technique to measure the attenuation of differential mode noise provided by common mode chokes. The proposed setup is a simpler alternative to the balanced setup commonly employed to that end, and its main advantage is that it avoids the use of auxiliary circuits (baluns). We make use of a modal analysis of a high-frequency circuit model of the common mode choke to identify the natural modes actually excited both in the standard balanced setup and in the simpler alternative setup proposed here. This analysis demonstrates that both setups are equivalent at low frequencies and makes it possible to identify the key differences between them at high frequencies. To analyze the scope and interest of the proposed measurement technique we have measured several commercial common mode chokes and we have thoroughly studied the sensitivity of the measurements taken with the proposed setup to electric and magnetic couplings. We have found that the proposed setup can be useful for quick assessment of the attenuation provided by a common mode choke for differential mode noise in a frequency range that encompasses the frequencies where most electromagnetic compatibility regulations impose limits to the conducted emissions of electronic equipment.

Keywords: electromagnetic compatibility; power electronics EMC; EMI mitigation techniques; EMI filter design and optimization

---

1. Introduction

The control of conducted emissions of electronic devices is an increasingly critical topic due to the trends toward the use of higher switching frequencies in power converters [1]. To mitigate conducted emissions, power line electromagnetic interference (EMI) filters are commonly used. However, at frequencies ranging from several hundred kilohertz to a few tens of megahertz the performance of EMI filters is typically undermined by parasitic effects such as parasitic parallel capacitances between windings in inductors and parasitic series inductances of capacitors [2–5]. Therefore, characterizing the response of these components at those high frequencies is becoming increasingly important to reduce design time, cost and size of the filter.

Common mode chokes (CMCs), made up of a pair of tightly coupled inductors, are key components of EMI filters primarily intended to limit common mode (CM) noise [2,5]. However, CMCs typically exhibit an inductive response to differential mode (DM) noise (leakage inductance) which has a significant impact on the attenuation that the EMI filter provides to differential mode (DM) noise [2]. The reason is that the leakage inductance provides a 40 dB/dec roll-off above its frequency of resonance with capacitors that typically are placed between power lines in the EMI filter to attenuate DM noise (Cx capacitors). This resonance typically occurs at tens or a few hundreds of kilohertz. At even higher frequencies (up to tens or a few hundreds of megahertz) leakage inductance no longer determines the DM attenuation of the CMC, which is instead governed by parasitic capacitive effects [6]. In this context, a method to directly measure and evaluate the actual frequency-dependent insertion loss provided by a CMC for DM noise would be very useful to estimate its suitability for a particular EMI filter.

A first option available to analyze the DM response of a CMC is to perform a full characterization of the device in a broad frequency range, obtaining an equivalent circuit which would make it possible to estimate the response of the CMC to a DM excitation by simulation as an additional task to perform after the characterization process is complete. Within this category of full characterization methods, several methods have been reported to obtain equivalent circuits for transformers and coupled coils [7–11]. Some of these methods have been particularized or directly conceived to be applied to CMCs [6,12,13]. The main drawback of these methods is that they often involve impedance measurements for different connections of the CMC and that they require a post-processing of the obtained data. Moreover, compensation is often required to perform these impedance measurements. Therefore, these methods are not the most appropriate option when a quick assessment of the DM response of a CMC is required.

A more straightforward alternative for evaluating the attenuation of DM signals provided by a CMC is to directly measure it. Techniques to measure the response of a four-port components such as CMCs to CM and DM excitations are adequately described in different standards, mostly based upon CISPR recommendations [14]. However, while measuring the CM attenuation of a CMC is fairly simple, the measurement of the insertion loss for a DM excitation requires the use of either a four-port vector network analyzer (VNA) or a balanced circuit that includes 180° splitters (baluns) [15]. Four-port VNAs are expensive and consequently may well not to be available. Baluns are much more affordable and common devices. However, its effect should be carefully assessed and taken into account because they may introduce some losses and, more importantly, they have a limited bandwidth [15]. Consequently, the use of these ancillary circuits complicates the setup and the measurement process. In this context, the availability of a simpler measurement technique that allows evaluation of the DM attenuation provided by a CMC would greatly facilitate the processes of design and test of EMI filters.

In response to that need, in this work we present a simple unbalanced measurement setup that can be used to quickly evaluate the attenuation that a CMC provides against DM noise. We perform a thorough analysis of this unbalanced setup which shows that possible sensibility of the unbalanced setup to external couplings might be a source of measurement errors. To determine whether this represent a problem in practice we systematically compare measurements performed with the unbalanced setup with those provided by the alternative balanced setup with the aim of clearly determining the validity, scope and limitations of the proposed setup.

This paper is organized as follows: in Section 2 we present a general analysis of the problem. In Section 2.1 we make use of a modal analysis of a high-frequency circuit model of a CMC considered to be a four-port network to obtain closed-form expressions for the transmission coefficients (and their corresponding frequencies of resonance) corresponding to both the standard balanced setup and the proposed unbalanced setup that can be used to assess the DM response of a CMC. The aim is to analyze the differences between both setups, and to identify the approximations under which both measurement methods provide similar results. Sections 2.2 and 2.3 present analysis of the effect of electric and magnetic couplings on the measurements performed with the balanced and unbalanced setups, with a focus on situations where these effects may arise in practice. In Section 3 we present results for several commercial CMCs to validate the analysis presented in Section 2. Also, in that section the actual sensitivity of both balanced and unbalanced setups to the effects of electric and magnetic couplings are experimentally studied. Finally, conclusions and a discussion about the scope of the method are provided in Section 4.

2. Analysis

Figure 1 shows a simplified representation of a CMC along with a lumped-element circuit model of the CMC. As shown in Figure 1a, a CMC is made up of two equal magnetically coupled windings. In [16] it has been demonstrated that a CMC can be conveniently modeled in a sufficiently broad frequency range by using a lumped-elements circuit with two blocks, each one containing two perfectly coupled inductances as shown in Figure 1b. The first block in that figure (CM block) only affects the CM noise, while the second block (DM block) contains inductors with opposite (perfect) coupling and therefore it only affects DM signals. In that circuit model parasitic intra-winding capacitances (C_t) and inter-winding capacitances (C_w) have been added to account for the response of the CMC throughout a sufficiently broad frequency range [16,17]. Also, losses within the magnetic material are accounted for in that model by resistors R_{CM} and R_{DM} placed in parallel with the coupled inductors. Finally, capacitances to ground, C_g, have been included in the circuit model to consider possible electric coupling to nearby metallic surfaces, e.g., the ground plane on a printed circuit board (PCB).

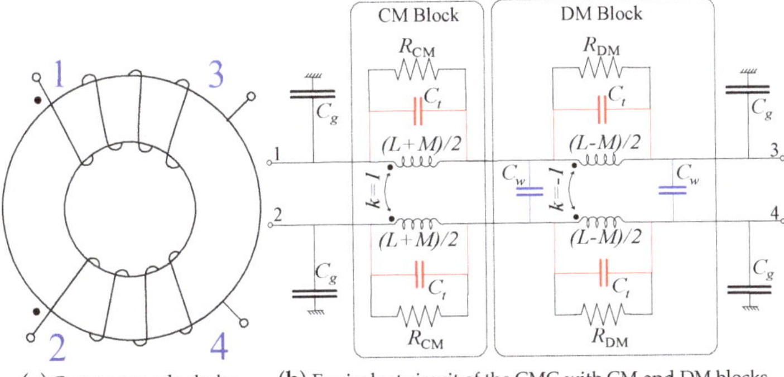

(**a**) Common mode choke. (**b**) Equivalent circuit of the CMC with CM and DM blocks.

Figure 1. Representation and circuit model of a common mode choke made up of two equal coupled windings with self-inductance L and mutual inductance M.

2.1. Modal Analysis of the CMC

The circuit in Figure 1b can be considered to be a four-port network and it can be characterized by a 4×4 admittance matrix $[Y]$. A modal analysis can be carried out by calculating the voltage eigenvectors (modes) that diagonalize $[Y]$, as explained in [16]. In that work, it has been shown that this analysis yields four independent (uncoupled) modes, which are referred to as C, V, H and D modes. In general, for a given excitation of the CMC, its response is always made up of a superposition of the responses of one or more of those natural modes. The equivalent circuits of these four modes are represented in Figure 2. In that figure we also represent the normalized excitation at the four ports of the CMC that corresponds to each mode. Please note that CM and DM excitation of the CMC appear as H and D natural modes of the equivalent circuit of the CMC (Figure 2c,d). As for the other two modes, mode C in Figure 2a corresponds to applying a common voltage to the four ports of the CMC, while mode V in Figure 2b is obtained by applying a difference of voltage between the two windings of the CMC.

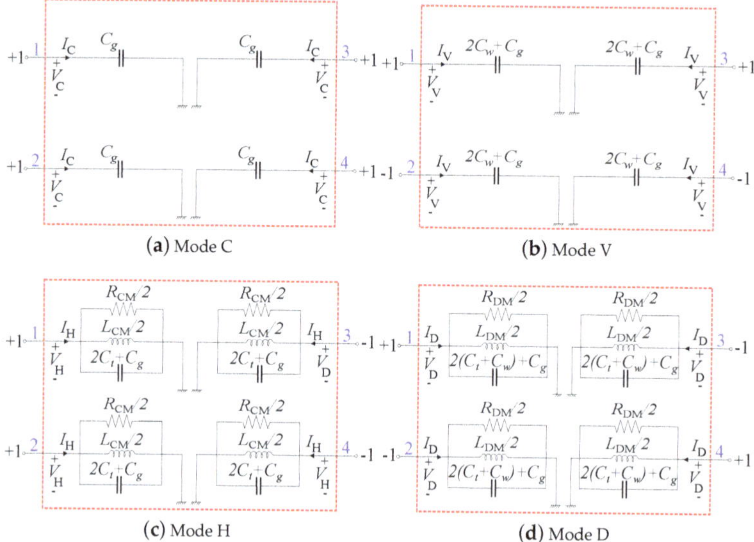

Figure 2. Equivalent circuits of the modes obtained for the high-frequency circuit of the CMC in Figure 1b. Normalized voltages at the four terminals of the CMC are indicated for each mode.

In Figure 2 it can be seen that the admittances of the four modes of the CMC are either a capacitive admittance (modes C and V) or the admittance of a resistor, an inductor and a capacitor connected in parallel (parallel RLC circuit), where the inductive component is proportional to $L_{\mathrm{CM}} = L + M$ for the CM (H mode) and to $L_{\mathrm{DM}} = L - M$ (the leakage inductance) for the DM (D mode). The admittances of the four modes of the CMC can be written as follows:

$$Y_{\mathrm{C}} = j\omega C_g. \tag{1}$$

$$Y_{\mathrm{V}} = j\omega(2C_w + C_g). \tag{2}$$

$$Y_{\mathrm{H}} = j\omega(2C_t + C_g) + \frac{2}{j\omega L_{\mathrm{CM}}} + \frac{2}{R_{\mathrm{CM}}}. \tag{3}$$

$$Y_{\mathrm{D}} = j\omega(2C_t + 2C_w + C_g) + \frac{2}{j\omega L_{\mathrm{DM}}} + \frac{2}{R_{\mathrm{DM}}}. \tag{4}$$

The modal analysis outlined above allows performance of a very efficient analysis of the measurement setups that can be used to characterize the DM response of CMCs. The idea is to express the transmission coefficients corresponding to these measurement setups in terms of the admittances of the natural modes of the CMC to determine to what extent those setups actually excite a pure DM (as intended) and also to assess the impact on measurements of the parasitic effects incorporated to the equivalent circuit of the CMC.

The measurement setup usually required to characterize the DM response of a CMC is schematically shown in Figure 3a [14]. We will refer here to this setup as DM setup. The DM setup requires the use of two baluns (or 180° dividers) to convert the excitation of the output port of the measurement device into a DM signal and to measure the transmitted DM signal at the input port. In other words, the DM setup permits direct measurement of the S_{DD21} term of the Mixed-Mode S-Parameter Matrix of the CMC seen as a four-port network [18]. This S_{DD21} S-Parameter physically represents the DM response of the CMC to a DM excitation, or equivalently the transmission coefficient (or inverse of the insertion loss) of the DM. For this reason, we will refer here to this S-parameter as S_{21}^{DM}. Table 1 provides S_{21}^{DM} as a function of the admittances of the natural modes of the CMC.

Since only Y_D and Y_V appear in the expression of S_{21}^{DM}, we can conclude that only D (DM) and V modes of the CMC are excited in the DM setup. Also, from that expression it is possible to obtain the frequency of resonance of the CMC in that setup in terms of the elements of the circuit model of the CMC in Figure 1b. This frequency of resonance, f_{DM}, is also given in Table 1 (Since in practice $R_{CM}, R_{DM} \gg R = 50\,\Omega$ the frequency of resonance can be calculated by taking $R_{CM} \to \infty$ and $R_{DM} \to \infty$ in S_{21}^{DM} and imposing $S_{21}^{DM} = 0$). At frequencies below f_{DM} the response of the CMC is dominated by the inductive part of Y_D, i.e., L_{DM}. Above f_{DM} the CMC behaves capacitively and the magnitude of S_{21}^{DM} increases with frequency. Please note that since f_{DM} does not depend on C_g, an electric coupling to ground will not alter the frequency of resonance of S_{21}^{DM}.

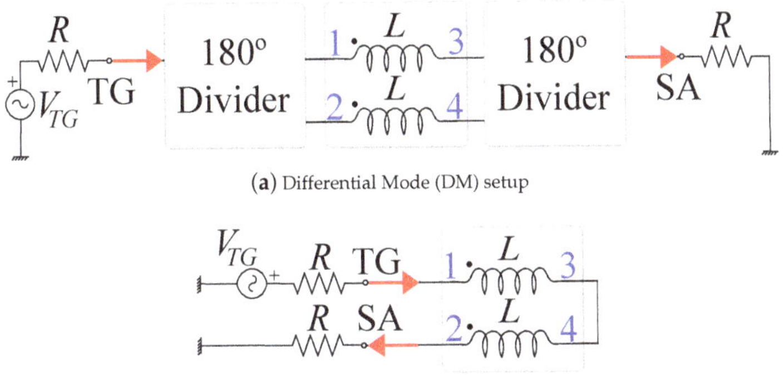

(**a**) Differential Mode (DM) setup

(**b**) Unbalanced Differential Mode (UDM) setup

Figure 3. Balanced and unbalanced setups for measuring transmission coefficients conveying information about the attenuation provided by a CMC for a differential mode excitation. Measurements can be performed with a spectrum analyzer (SA) with tracking generator (TG) or a VNA.

Table 1. Transmission coefficients and frequencies of resonance for a CMC measured in the setups of Figure 3, where $Y_{OC} = Y_H Y_D/(Y_H + Y_D)$. Approximated expressions assume $C_g \ll C_t, C_w$ and $Y_C \ll Y_V, Y_H, Y_D$.

Setup	Transmission Coefficient	Frequencies of Resonance
DM	$S_{21}^{DM} = \frac{RY_D}{2+RY_D} - \frac{RY_V}{2+RY_V}$	$f_{DM} = \frac{1/2\pi}{\sqrt{C_t L_{DM}}}$
UDM	$S_{21}^{UDM} \approx \frac{R(Y_D+Y_V)}{2+R(Y_D+Y_V)}$	$f_{UDM} \approx \frac{1/2\pi}{\sqrt{(C_t+2C_w)L_{DM}}}$
OC	$S_{21}^{OC} \approx \frac{2RY_{OC}}{2RY_{OC}+1}$	$f_{OC} \approx \frac{1/2\pi}{\sqrt{(C_t+C_w)L_{DM}}}$

As an alternative to the DM setup, in Figure 3b we propose a simpler unbalanced setup (UDM setup) which dispenses with baluns and which at the same time is also able to provide information about the DM response of the CMC. The UDM setup in Figure 3b, like the DM setup in Figure 3a, involves the measurement of a transmission coefficient instead of the measurement of impedances of the CMC. This allows for avoiding additional measurements to account for the effect of cables and/or test fixtures (compensation measurements) [19]. To compare the UDM setup with the DM setup, it is useful to obtain also the transmission coefficient of the UDM setup (S_{21}^{UDM}) in terms of the admittances of the natural modes of the CMC. An analysis of the circuit in Figure 3b with the circuit model of the CMC in Figure 1b leads to:

$$S_{21}^{UDM} = \frac{R(Y_p - 4Y_s)}{(2 + Y_p R)(1 + 2Y_s R)}. \tag{5}$$

where $Y_p = Y_D + Y_V + 2Y_C$ and $Y_s = Y_C Y_H/(Y_C + Y_H)$. This analysis shows that unlike the DM setup which only excites D and V modes, the UDM setup excites the four natural modes of the CMC. In principle, this makes an important difference between DM and UDM setups. However, it can be easily shown that if measurements are performed avoiding the presence of nearby conducting surfaces, the response of the UDM setup becomes quite similar to that of the DM setup. In particular, only D and V modes are significantly excited in the UDM setup. To demonstrate this, suppose that capacitances to ground are negligible compared with the rest of the capacitances of the circuit model in Figure 1b ($C_g \ll C_t, C_w$). In that case, the expressions for the modal impedances in Equations (1)–(4) allow us to assume that $Y_\mathrm{C} \ll Y_\mathrm{V}, Y_\mathrm{H}, Y_\mathrm{D}$. Consequently, we can approximate in (5) $Y_s \approx Y_\mathrm{C} \ll Y_p \approx Y_\mathrm{D} + Y_\mathrm{V}$. In that case, a simpler approximation of the expression of S_{21}^{UDM} in (5) can be obtained:

$$S_{21}^{\mathrm{UDM}} \approx \frac{R(Y_\mathrm{D} + Y_\mathrm{V})}{2 + R(Y_\mathrm{D} + Y_\mathrm{V})}. \tag{6}$$

This expression is included in Table 1 along with that of S_{21}^{DM}. The frequency of resonance of S_{21}^{UDM} in Equation (6) is also provided in that table as f_{UDM}. When comparing S_{21}^{UDM} with S_{21}^{DM}, it can be noticed that both expressions depend only on Y_D and Y_V. Moreover, because Y_V in Equation (2) is a capacitive admittance related to the inter-winding capacitance, it is expected that at low frequencies $Y_\mathrm{V} \ll Y_\mathrm{D}$ and therefore $S_{21}^{\mathrm{UDM}} \approx S_{21}^{\mathrm{DM}} \approx RY_\mathrm{D}/(2 + RY_\mathrm{D})$. In other words, at low frequencies (i.e., well below resonance) the response of the CMC is expected to be dominated by L_{DM} and to be the same for the DM setup as for the UDM setup, with currents flowing inside the CMC in a purely differential mode in both cases.

Summing up, the previous analysis demonstrates that provided that coupling with nearby metallic surfaces is negligible, $S_{21}^{\mathrm{UDM}} \approx S_{21}^{\mathrm{DM}}$ at low frequencies. This is an important result because it allows us to conclude that at these low frequencies the UDM setup can be used as a simpler alternative to the DM setup to quickly characterize the DM response of a CMC. However, the analysis presented here also demonstrates that at high frequencies some differences should be expected between S_{21}^{DM} and S_{21}^{UDM}. This will be experimentally checked in Section 3. Before that the next subsections complete the theoretical analysis by analyzing the effect of electric and magnetic couplings of the CMC with nearby conducting surfaces on measurements performed with DM and UDM setups.

2.2. Effect of Electric Coupling to Metallic Surfaces

In the previous section we have seen that the transmission coefficient S_{21}^{DM} of the balanced DM setup is inherently independent of Y_C (see Table 1) and that electric couplings are not expected to significantly affect measurements carried out with the DM setup. By contrast, external electric couplings will affect measurements with the UDM setup unless S_{21}^{UDM} can be approximated by Equation (6). That approximation can be safely applied whenever no metallic surface is allowed near the CMC when measuring. However, there are situations where this cannot be ensured. This may occur for example when measuring a CMC which is mounted on a PCB with a return plane. This PCB can be for example the PCB of the EMI filter or a PCB employed to measure the CMC as a four-port terminal, as required in many cases in the normative [14].

Even for a CMC mounted on a PCB, the return plane is usually sufficiently far from the windings of the CMC as to make it reasonable to disregard a direct electric coupling of the CMC to the return plane with respect to the internal couplings given by the inter-windings and intra-windings capacitances ($C_g \ll C_t, C_w$ in Figure 1b). However, the effects on the DM and UDM measurements of the microstrip traces of the PCB employed to lead the signals to the four pins of the CMC should be assessed. As for the DM measurements, it can be easily demonstrated that if the characteristic impedance of these signal traces of the PCB are equal to the input and output impedances of the measuring device (usually 50 Ω), those traces will only introduce a phase shift in the transmission coefficient of the DM setup, but they will not alter the magnitude of S_{21}^{DM} [19]. However, the situation is different for the UDM setup.

A schematic of the situation presented for the UDM setup can be seen in Figure 4. That figure represents a CMC connected in the UDM setup, and it includes transmission line models for the interconnecting cables (e.g., coaxial cables), which are labelled as $\mathrm{TL_C}$ lines, and for the signal traces of the PCB (typically microstrip lines), which are labelled as $\mathrm{TL_T}$ lines. In principle, $\mathrm{TL_T}$ traces leading to terminals 1 and 3 of the CMC are not an issue. In fact, provided that they have the same 50 Ω characteristic impedance as TL_C lines, those traces will only introduce a phase shift in the transmission coefficient [19]. However, TL_T lines attached at terminals 2 and 4 of the CMC create an asymmetric electric coupling of the CMC to ground whose effect is not negligible, as we will show here. To demonstrate that, consider the situation represented in Figure 4, where terminals 2 and 4 of the CMC are directly short-circuited to achieve an UDM configuration that circumvents the two TL_T lines attached to those terminals. Considering lengths of TL_T lines in the order or centimeters, they will be electrically short at the frequencies of interest and, consequently, these open-circuited lines can be modeled as capacitances [20]. The two parasitic capacitances C_r included in Figure 4 account for this effect. Since these C_r capacitances correspond to the total capacitance of the trace lines, their values are typically in the order of units or tens of picofarads [20]. Consequently, C_r capacitances cannot be disregarded by comparison with typical parasitic capacitances of the CMC, which are of the same order of magnitude.

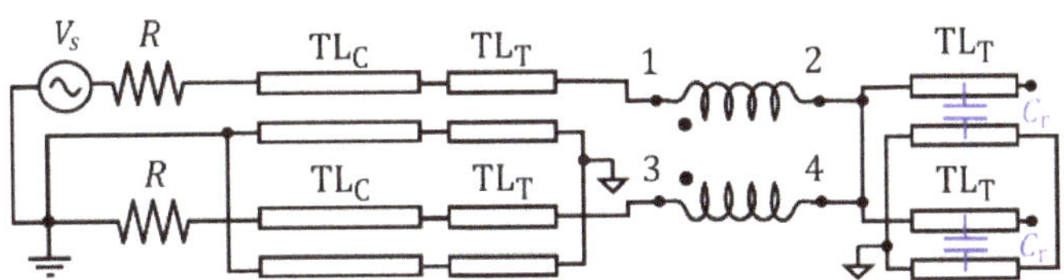

Figure 4. Schematic of a CMC mounted on a grounded PCB connected in UDM setup. Transmission lines labelled as TL_C stand for the interconnecting cables. Transmission lines labelled as TL_T represent the signal traces of the PCB. The signal traces terminated as open-circuits at terminals 2 and 4 are supposed to be electrically short and thereby modeled as two capacitances C_r to ground.

To investigate the impact on the transmission coefficient measured with the UDM setup of the capacitances C_r that must be included in the schematic of the UDM setup in Figure 4, we have calculated the transmission coefficient of that circuit by circuit analysis. We have used the circuit model of the CMC in Figure 1 with $Y_C = 0$, but we have included the effect of C_r capacitors as two external admittances $Y_R = j\omega C_r$ connected to terminals 2 and 4 of the CMC. In this way, we have obtained the following modified S_{21}^{UDM} coefficient in terms of the admittances of the modes of the CMC:

$$S_{21}^{\mathrm{UDM}'} = \frac{R(Y_D + Y_V)}{2 + R(Y_D + Y_V)} - \frac{RY_{HR}}{2 + RY_{HR}}. \tag{7}$$

where $Y_{HR} = Y_H Y_R / (Y_H + Y_R)$. It is very interesting to note that the only difference between $S_{21}^{\mathrm{UDM}'}$ in (7) and S_{21}^{UDM} in Equation (6) is the presence of the second additive term in (7). This term is zero if $Y_R = 0$ (C_r=0), as expected. Therefore, Equation (7) reveals that the effect of considering the electric coupling given by C_r is to excite an additional mode of the CMC (H mode), whose admittance Y_H acts in series with that of C_r. A very good approximation for the frequency of resonance of $S_{21}^{\mathrm{UDM}'}$ in Equation (7) can be obtained if we consider that this frequency must be near that of S_{21}^{UDM}, i.e., it must be found at frequencies where $\omega C_t \approx 1/|\omega L_{DM}|$. Because in practical CMCs we have $L_{CM} \gg L_{DM}$, we have $1/|\omega L_{CM}| \ll \omega C_t$ and therefore Y_H in Equation (3) can be approximated as $Y_H \approx j\omega C_t$. With this approximation the frequency of resonance of $S_{21}^{\mathrm{UDM}'}$ in Equation (7) can be expressed as:

$$f'_{\mathrm{UDM}} = \frac{1/2\pi}{\sqrt{(C_t + 2C_w - \frac{C_r C_t}{C_r + C_t})L_{\mathrm{DM}}}}. \tag{8}$$

By comparing this frequency of resonance with f_{UDM} given in Table 1 we conclude that the main effect of C_r in the transmission coefficient of the UDM setup is to slightly increase its frequency of resonance.

From the previous analysis we conclude that by contrast with DM setup, the UDM setup is sensitive to the presence of electrical couplings (capacitive effects) caused by the presence of nearby metallic grounded structures such as a ground plane on a PCB, and that this effect can be noticeable at high frequencies.

An interesting question that arises in this point is whether in the UDM setup this effect can be avoided by isolating the return plane of the PCB with respect to ground in the measurement setup. This situation is shown in Figure 5, which represents an alternative implementation of the UDM setup for a CMC mounted on a PCB where the active conductors of the TL_C lines (cables) are directly connected to the terminals of the CMC, while circumventing the TL_T lines connected to terminals 1 and 3, thus isolating the return plane of the PCB from the ground of the measurement setup. In that case, the four electrically short open-circuited TL_T lines connected at each terminal will introduce four parasitic capacitances C_r at each terminal of the CMC.

These capacitances connected to a star point (the isolated return plane) can be transformed into its triangle equivalent, as shown in figure Figure 6. That figure shows that the effect of the capacitances C_r is to increase the C_w parasitic capacitance of the CMC by an amount $\Delta C_w = C_r/4$, and to increase C_t by an amount $\Delta C_t = C_r/2$. This should cause a decrease in the frequency of resonance of S_{21}^{UDM} with respect to the case with no return plane. Therefore, the UDM Setup will still be sensitive to the presence of the return plane of the PCB. This will be experimentally verified in Section 3.

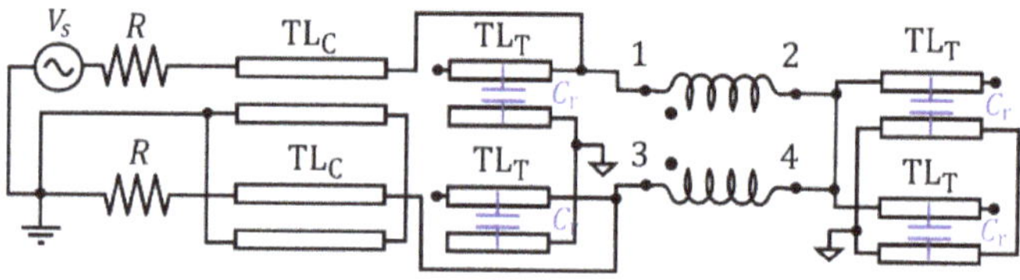

Figure 5. Equivalent circuit of the UDM setup for a CMC mounted on a PCB with an ungrounded return plane. Transmission lines labelled as TL_C stand for the interconnecting cables. Transmission lines labelled as TL_T represent the signal traces of the PCB. Since all the signal traces are terminated as open-circuits and they are supposed to be electrically short, they are actually modeled as capacitances C_r to ground.

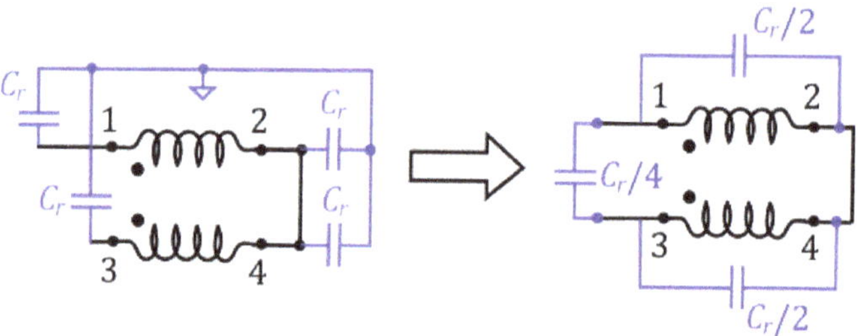

Figure 6. Star to triangle conversion for the parasitic capacitances C_r that appear at the four terminals of the CMC in the circuit of Figure 5.

2.3. *Effect of Magnetic Coupling to Metallic Surfaces*

Another effect that can alter the response of a CMC excited by a DM signal is a magnetic coupling of the CMC with nearby metallic surfaces. The root cause of this effect is that, unlike CM currents, DM currents in a CMC create a magnetic field which closes its field lines outside the core of the CMC [21,22]. Therefore, this magnetic field can interact with nearby metallic surfaces such as for instance, the metallic enclosure usually employed for housing and shielding of EMI filters. Even though metallic enclosures are usually constructed with non-magnetic metals, a magnetic coupling might still appear due to eddy currents induced in those conducting surfaces by the time-varying stray magnetic field of the CMC. The effect of these eddy currents is to partially counteract the magnetic fields created by the CMC outside its core, thus causing a decrease of the leakage inductance L_{DM} of the CMC [22,23]. This change of L_{DM} can equally affect measurements performed with the DM or the UDM setups. The actual impact of this effect in practical cases will be investigated in Section 3.3.

3. Results

The analysis presented in the previous section suggests that in principle it is possible to use the UDM setup in Figure 3b to measure the DM response of a CMC. However, since we have shown that electric or magnetic couplings may affect the response of the CMC, the impact of these effects on measurements performed with the UDM setup must be assessed to clearly determine the scope and limitations of this method of measurement by comparison with standard DM measurement. To this end, we present in this section experimental results for several commercial CMCs in different setups. These measurements have been carried out by using a Rhode&Schwarz ZND VNA. However, note that measurements of the magnitude of a transmission coefficient can be alternatively performed by using a spectrum analyzer with tracking generator.

3.1. Analysis of the Response of a Standalone CMC

To validate the analysis presented in Section 2 we have measured the response of the CMCs listed in Table 2 in the DM and UDM setups shown in Figure 3. Also, we have used an alternative setup, referred to as Open-Circuit (OC) setup, which is shown in Figure 7. This setup was proposed in [16] to characterize CMCs. The main difference between OC setup and UDM is that since the OC setup is actually used for obtaining a complete circuit model of the CMC and it is not conceived as a measurement method to characterize its DM response, in the OC setup both the CM (H) and DM (D) modes of the CMC are simultaneously excited. An expression for the transmission coefficient of the CMC in the OC setup, S_{21}^{OC} in terms of Y_H and Y_D is given in Table 1 [16]. From this expression, it can be easily demonstrated that S_{21}^{OC} always presents two frequencies of resonance: one related to L_{CM} and another one at a higher frequency associated with L_{DM} which is given in Table 1 as f_{OC}. As a consequence of this double resonance, the OC setup does not allow quick measurement of the response of a CMC to a purely DM signal throughout the entire range of frequencies of interest. However, since the expression and physical meaning of f_{OC} has been previously studied in [16], we will use it here to validate our analysis of the UDM setup as explained below.

Table 2. Parameters extracted for the equivalent circuit of Figure 1b for several commercial common mode chokes.

Manufacturer and Part Number	L (mH)	L_{CM} (mH)	L_{DM} (uH)	C_w (pF)	C_t (pF)	R_{CM}(kΩ)	R_{DM}(kΩ)
WÜRTH ELEKT. 744824622	2.2	4.94	4.7	4.2	6.8	17.2	6.5
WÜRTH ELEKT. 744824310	10	26.7	33.6	4.7	18.3	118	16.6
WÜRTH ELEKT. 744824220	20	54.1	57.6	10.7	20.2	203	22.2
WÜRTH ELEKT. 7448011008	8.0	6.90	6.5	0.86	2.7	22.1	8.1
MURATA PLA10AN2230R4D2B	22	71.3	173	1.8	2.9	73.9	33.0
KEMET SC-02-30G	3.0	7.40	5.8	1.4	2.8	34.3	16.7
KEMET SCF20-05-1100	11	13.4	5.1	8.6	7.2	15.4	7.9

By comparing the frequencies of resonance in Table 1 it is apparent that for a given CMC we must have $f_{UDM} < f_{OC} < f_{DM}$. We will use here this fact to verify the accuracy of our circuit model and to check the expressions for the transmission coefficients of the DM and UDM setups, S_{21}^{DM} and S_{21}^{UDM}, presented in Table 1.

Figure 8 shows $|S_{21}^{DM}|$, $|S_{21}^{UDM}|$ and $|S_{21}^{OC}|$ measured for the CMC listed as WÜRTH ELEKTRONIK 744824622 (2.2 mH) in Table 2. It can be observed that each one of these curves present a resonance dip related to differential excitation of the CMC, as expected. Also, note that these frequencies of resonance are different for the three setups. In fact, resonance occurs first for the UDM setup, then for the OC setup (second resonance) and finally, for the DM, i.e., $f_{UDM} < f_{OC} < f_{DM}$. These results are consistent with our previous analysis. Similar results are obtained for all the CMCs in Table 2. An additional example is represented in Figure 9 for another CMC, identified in the caption of the figure and listed in Table 2.

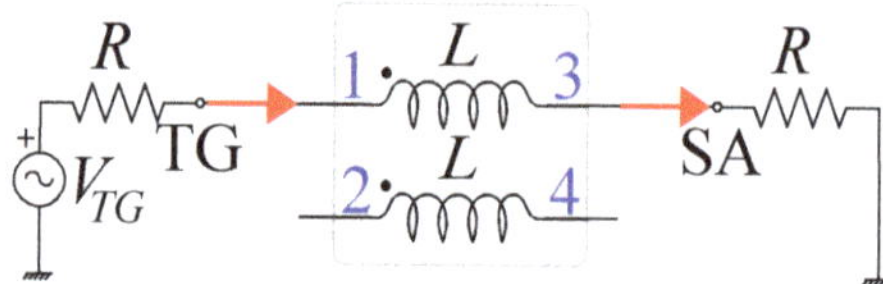

Figure 7. Open-circuit (OC) setup proposed in [16] for characterizing CMCs.

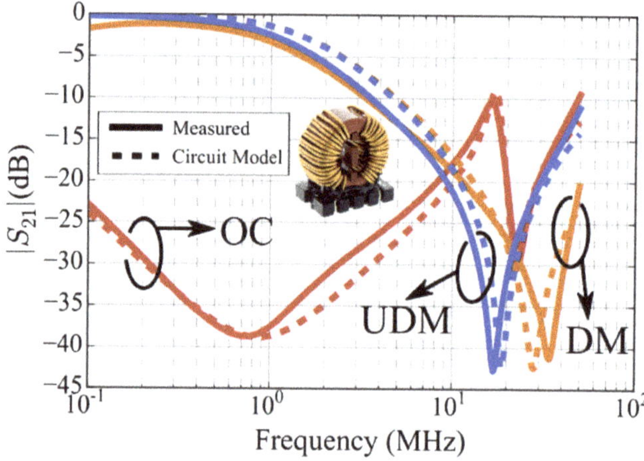

Figure 8. Magnitude $|S_{21}|$ for the CMC listed as Würth Elecktronik 744824622 (2.2 mH) in Table 2, using the setups in Figures 3 and 7.

To further ensure consistency of the measured curves with our theoretical analysis, we have used an advanced search algorithm based upon genetic algorithms (GA) [24] to find a set of values for the components of the circuit in Figure 1b that allow us to simultaneously fit the measured S_{21}^{OC} and S_{21}^{DM} curves. Parameters obtained for all the CMCs analyzed here are given in Table 2. Then, we have used these circuit parameters to calculate $|S_{21}^{UDM}|$ using the expression given in Table 1. These curves are represented in Figures 8 and 9 as dashed lines, and labelled as calculated results. A good agreement between measured and calculated results can be observed in these graphs. Similar results are obtained for all the CMCs in Table 2. This permits us to ensure that the high-frequency circuit model in Figure 1b is reasonably accurate within the range of frequencies where most EMC regulations impose limits to conducted emissions [25–27].

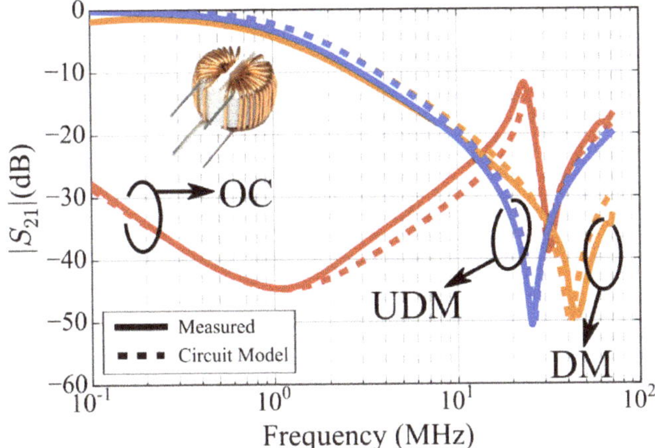

Figure 9. Magnitude $|S_{21}|$ for the CMC listed as KEMET SC-02-30G (3 mH) in Table 2, using the setups in Figures 3 and 7.

It is worth pointing out that in the measurements performed in this section external magnetic or electric couplings have been carefully avoided by keeping the CMC under test far from metallic surfaces (a distance greater than the size of the CMC is typically enough). This is important because the expression for S_{21}^{OC} in Table 1, such as that of S_{21}^{UDM} Equation (6), assumes that parasitic capacitances to ground can be disregarded, i.e., $Y_C \ll Y_V, Y_H, Y_D$. The effect on measurements of electric coupling with nearby metallic surfaces will be analyzed in the next subsection.

3.2. Effect of Capacitive Couplings in a PCB

The analysis presented in Section 2.2 shows that when a CMC is mounted on a PCB (a situation that may easily arise in practice), the parasitic capacitances introduced by the traces connected to the terminals of the CMC might alter the measurements of S_{21}^{UDM}, thus rendering misleading results if the aim is to characterize the CMC as a standalone component. In this section, we will analyze the actual impact of this effect by comparing measured S_{21}^{UDM} for an isolated CMC with results obtained when the CMC is mounted on a PCB representative of those usually employed to fabricate EMI filters.

Figure 10 shows a CMC mounted on a PCB fabricated with a 1.5 mm-thick FR4 substrate. Signal traces (not visible in Figure 10 because the CMC is mounted on the side of the return plane) are 4 mm-width×60 mm-long strips connecting the CMC pins to the BNC connectors. These signal traces, along with the BNC connectors, add an extra parasitic capacitance between the four ports of the CMC and the return plane (RP). We have measured this parasitic capacitance, obtaining a value of $C_r = 17.5$ pF.

As a first step, we have checked that the fact that the CMC is mounted on a PCB has a negligible effect on the measurements of $|S_{21}^{DM}|$, as expected from the analysis in Section 2. This robustness of S_{21}^{DM} measurements against external electric couplings comes from the balanced nature of this setup, and represents an advantage of this measurement technique. On the contrary, and according to the analysis in Section 2.2, the measure of $|S_{21}^{UDM}|$ could be affected by the parasitic capacitances appearing at the terminals of the CMC when it is mounted on a PCB with a RP. To study the actual impact of this effect in a practical case, Figure 11 shows $|S_{21}^{UDM}|$ measured for the CMC listed as WÜRTH ELEKTRONIK 744824220 (20 mH) in Table 2 in three different situations: when CMC is isolated (no magnetic or electric coupling with nearby conducting surfaces), when CMC is mounted in the PCB but the RP is not grounded (schematic in Figure 4) and finally, when the CMC is mounted in the PCB and the RP is grounded (schematic in Figure 5). Curves in Figure 11 show that the response of the CMC at low frequencies (well below resonance) is the same for these three situations. However, Figure 11 also reveals that at high frequencies there exists a shift in the frequency of resonance of S_{21}^{UDM}. Compared to the isolated case, the frequency of resonance is slightly lower when the RP is not grounded and slightly higher when the RP is grounded. These results are consistent with the analysis presented in Section 2.2. We have obtained similar results for all the CMCs in Table 2.

Figure 10. A 2.2 mH CMC, listed in Table 2 as WÜRTH ELEKTRONIK 744824622, mounted on a PCB fabricated to check the impact on $|S_{21}^{\mathrm{UDM}}|$ of the capacitive coupling of the signal traces to the return plane.

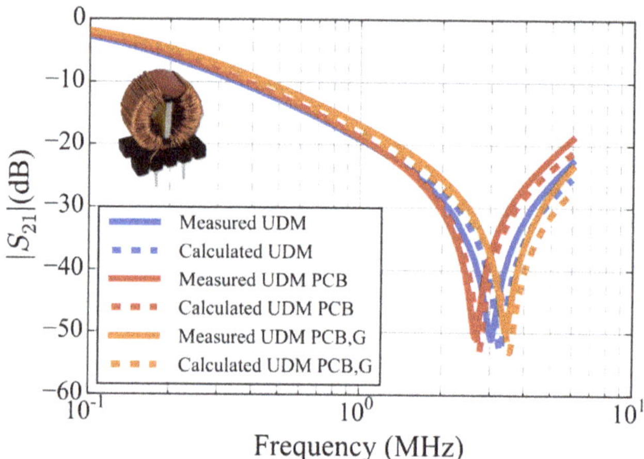

Figure 11. Measured and calculated $|S_{21}^{\mathrm{UDM}}|$ curves for the CMC listed as WÜRTH ELEKTRONIK 744824220 (20 mH) in Table 2. We compare curves for three cases: standalone CMC, CMC mounted on a PCB with a floating return plane and CMC mounted on a PCB whose return plane is grounded (G label).

To analyze this effect in more detail, and also to check the explanation provided inSsection 2.2, we have measured the frequencies of resonance of three different CMCs (among those in Table 2) when measured in the three different situations described above. In Table 3 we compare the frequency of resonance measured with the CMC isolated with that obtained when the CMC is placed on a PCB with an ungrounded RP. Those results show that the effect of the presence of the isolated RP is to decrease the frequency of resonance of S_{21}^{UDM} by an amount that goes from 10% to 20%. Table 3 also includes for each CMC the frequency of resonance calculated by using the expression of f_{UDM} given in Table 1, where C_t and C_w have been modified to $C_t' = C_t + C_r/2$ and $C_w' = C_w + C_r/4$ in accordance with the results of the analysis presented in Section 2.2. The rest of the parameters of the model of each CMC have been taken from Table 2. Results in Table 3 show that the calculated frequency of resonance agrees reasonably well with the measured one, with typical discrepancies around 5%. This allows us to conclude that the decrease of the frequency of resonance of S_{21}^{UDM} observed when the CMC is mounted on a PCB with an isolated RP is mainly caused by the parasitic capacitances between the traces and the isolated return plane of the PCB.

Table 3. Measured frequencies of resonance of S_{21}^{UDM} for three CMCs when isolated and when mounted on a PCB with an ungrounded return plane. Also, calculated f_{UDM} for the latter case.

	f_{UDM} (MHz)		
	CMC Isolated	**CMC on Ungrounded PCB**	
CMC Part Number	**Measured**	**Measured**	**Calculated**
WE 744824622	16.9	13.6	12.8
WE 744824310	4.77	4.30	4.07
WE 744824220	3.02	2.65	2.74

Table 4 compares the frequency of resonance of S_{21}^{UDM} measured when the CMC is isolated with that measured when the CMC is mounted on a grounded PCB for the same three CMCs listed in Table 3. Results show that the grounded RP causes the frequency of resonance to increase in all the cases, as expected. This increase is up to 35% for the WE 2.2 mH CMC. Frequencies of resonance f'_{UDM} calculated by using (8) are also included in Table 4. The good agreement found in general between calculated and measured results confirms that the shift of the frequency of resonance of S_{21}^{UDM} is caused by the presence of the grounded RP and that this shift can be approximately predicted by using Equation (8).

Table 4. Measured frequencies of resonance of S_{21}^{UDM} for three CMCs when isolated and when mounted on a PCB with a grounded return plane. Also, calculated f_{UDM} for the latter case.

	f_{UDM} (MHz)		
	CMC Isolated	**CMC on Grounded PCB**	
CMC Part Number	**Measured**	**Measured**	**Calculated**
WE 744824622	16.9	21.7	22.8
WE 744824310	4.77	5.63	5.64
WE 744824220	3.02	3.40	3.72

Summing up, results in this section show that when measured in the UDM setup, the capacitance between the signal traces and the RP of the PCB can modify the response of the CMC at high frequencies. Therefore, when accurate results are required, measuring a CMC mounted on a PCB with the UDM setup should be avoided unless the effect of the parasitic capacitances of the signal traces to the RP is accounted for.

Another interesting conclusion that can be obtained from these results refers to the impact of magnetic coupling. Figure 11 shows that at low frequencies the response of the standalone CMC coincides with those measured when the CMC is mounted on a PCB. Therefore, we can conclude that at low frequencies (i.e., well below resonance), the RP has no significant effect on L_{DM}. This is because the RP is not sufficiently close to the windings of the CMC. Situations where magnetic coupling with external metallic surfaces may have an impact on the measurements of S_{21}^{UDM} (and S_{21}^{DM}) will be analyzed in the next subsection.

3.3. *Effect of Magnetic Coupling to Nearby Conducting Surfaces*

The aim of this section is to study the sensitivity of the measurements of S_{21}^{DM} and S_{21}^{UDM} to the effect of magnetic coupling to nearby conducting surfaces (CS). We have verified for all the CMCs in Table 2 that the measurements of S_{21}^{DM} and S_{21}^{UDM} are altered by the presence of an isolated (i.e., not necessarily grounded) copper plate placed very close to the top of the CMC. This situation is representative of that that may arise in practice when the CMC of an EMI filter is placed very close to the metallic enclosure of the filter. We have verified that this effect is noticeable only if the CS is very close to the windings (in general, less than 1 mm apart). In fact, we have checked that the effect of magnetic coupling is very weak when the CS is placed beneath (instead of on top) of the CMC, because in this case the leads and the plastic structure that typically supports the CMC keep the CS sufficiently far from the windings of the CMC to prevent a significant magnetic coupling. This is consistent with results for CMCs mounted on a PCB presented in the previous section, where we have seen that the effect of the return plane of the PCB on the leakage inductance of the CMC, L_{DM}, is negligible.

As an example to show the effect of magnetic coupling on the DM response of CMCs, we represent in Figure 12 the measured $|S_{21}^{UDM}|$ and $|S_{21}^{DM}|$ with and without the presence of a CS for the 10 mH CMC listed as WÜRTH ELEKTRONIK 744824310 in Table 2. The CS employed to induce magnetic coupling in this experiment is a 10cm-side square copper plate which has been placed on top of the CMC, only separated by a paper film whose thickness is approximately 0.1 mm. Figure 12 shows that the effect of the CS is to slightly increase both $|S_{21}^{DM}|$ and $|S_{21}^{UDM}|$ at low frequencies and also to increase their respective frequencies of resonance. These effects can be explained by a decrease of L_{DM}. To demonstrate this, we have included in Figure 12 the $|S_{21}^{DM}|$ and $|S_{21}^{UDM}|$ curves calculated by using the equivalent circuit of Figure 1b with the parameters extracted in the previous sections (Table 2) and also the same curves obtained after conveniently decreasing L_{DM}. Those calculated results agree very well with measurements. In general, the effect of the magnetic coupling created by CS placed very close to a CMC is to decrease the leakage inductance of the CMC by an amount between a 20% and a 30%. Detailed quantitative results are provided in Table 5 for the CMC in Figure 12 and two additional CMCs among those listed in Table 2. Table 5 compares for each CMC the inductance L_{DM} calculated without the presence of a CS with the reduced inductance (referred to as L_{DM}^{CS}) that should be used to match the S_{21}^{DM} and S_{21}^{UDM} curves in the presence of a CS. It is interesting to highlight that the same reduced inductance L_{DM}^{CS} can be used to account for the effect of the magnetic coupling for both the DM and UDM measurements, as should be expected.

The main conclusion that can be drawn from the experiment described in this section is that a closely placed CS can alter the DM response of a CMC by decreasing its leakage inductance. Moreover, this effect can equally affect measurements in both the DM and the UDM setups. This implies that independently of the setup, magnetic coupling should be avoided (by keeping the CMC sufficiently far apart from metallic surfaces) to prevent an inaccurate characterization of the DM response of a CMC as a standalone component. Alternatively, in case the actual metallic enclosure where the CMC is going to be placed is available, it would be possible to determine its effect on the leakage inductance of a CMC by measuring it in the UDM setup with the CMC placed in the same relative position with respect to the metallic enclosure that it is intended to occupy in practice.

Table 5. L_{DM} for different CMCs with and without the effect of a nearby conducting surface.

CMC Part Number	L_{DM} (μH)	L_{DM}^{CS} (μH)
WE 744824622	4.73	3.38
WE 744824310	33.6	27.0
WE 744824220	58.2	44.2

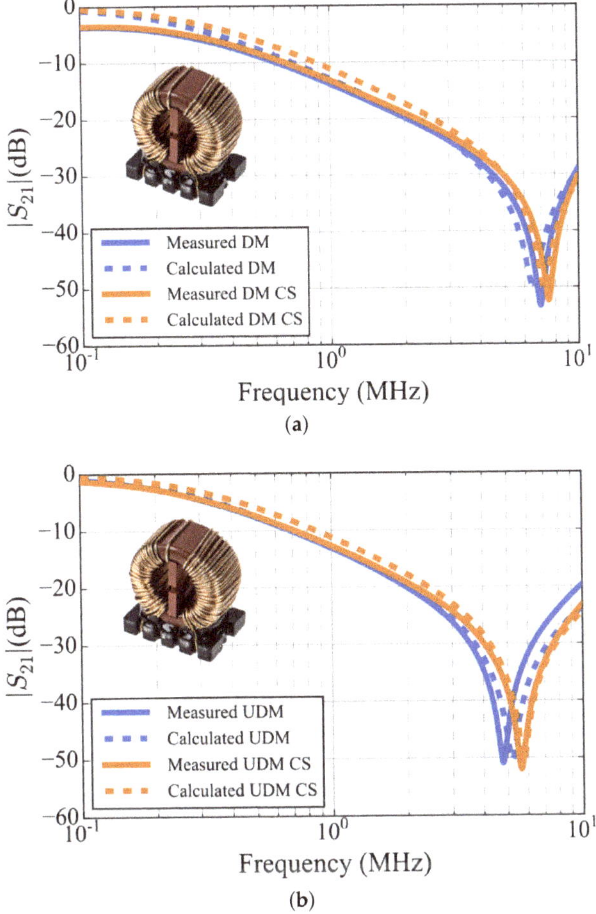

Figure 12. Measured and calculated $|S_{21}^{\mathrm{DM}}|$ (**a**) and $|S_{21}^{\mathrm{UDM}}|$ (**b**) for the CMC listed as WÜRTH ELEKTRONIK 744824310 (10 mH) in Table 2, whose picture is inserted in the figures. The CS acronym used in the legends indicates results corresponding to the case where a conducting surface is placed near the CMC.

From results in Figure 12, it is also interesting to note that below approximately 3 MHz, $|S_{21}^{\mathrm{DM}}|$ curves coincide very well with $|S_{21}^{\mathrm{UDM}}|$ ones. This is consistent with the discussion presented in Section 2.1, where we demonstrate that at low frequencies the response of a CMC is mainly determined by L_{DM} and that is expected that $S_{21}^{\mathrm{DM}} \approx S_{21}^{\mathrm{UDM}}$. This confirms that in the range of frequencies where the CMC behaves inductively, the UDM setup can be effectively used to predict the DM insertion loss of a CMC.

It is also worth pointing out that the curves for $|S_{21}^{\mathrm{DM}}|$ in Figure 12a show a slight attenuation at very low frequencies (between 100 kHz and 200 kHz) when compared with $|S_{21}^{\mathrm{UDM}}|$ curves in Figure 12b. This effect can also be observed in measured $|S_{21}^{\mathrm{DM}}|$ represented in Figures 8 and 9. We have verified that this is caused by a decrease in the performance of the baluns that we have employed in these measurements, due to its limited bandwidth (We have used baluns constructed with commercial wide-band 1:1 transformers Coilcraft WB2010-1 [28].). Although this problem could be solved by using baluns with a wider bandwidth or by performing a careful calibration, we point it out here to highlight an inherent shortcoming of the DM setup, namely its dependence on ancillary circuitry whose effect on the measurements has to be carefully taken into account.

4. Discussion and Conclusions

This work presented a thorough analysis of a measurement setup, referred to as UDM setup, which can be used to readily measure the response of a CMC to DM signals throughout the frequency range where most EMC regulations impose limits to conducted emissions. The UDM setup is conceived as a simpler and faster alternative to balanced setups, which require ancillary circuits (baluns), and to the use of sophisticated equipment such as four-port VNAs.

We have presented a detailed analysis of the UDM setup based on a modal analysis of a high-frequency circuit model of the CMC. This modal analysis has allowed us to obtain analytical expressions for the transmission coefficients of the CMC in terms of the admittances of the natural modes of the CMC, both for the DM and UDM setups. From these expressions it has been possible to determine the modes actually excited in the CMC in each setup which has permitted us to analyze the effect of parasitic effects on each setup and to identify the conditions that ensure similar responses of the CMC for the DM and UDM measurement setups.

The analysis and experimental results presented in this work provide a deep understanding on the differences and similarities between both techniques of measurement. This has allowed us to demonstrate that the DM setups is inherently immune to (symmetric) external electric coupling while the UDM is sensitive to this effect. By contrast, we have verified that both DM and UDM setups are equally sensitive to magnetic coupling of the CMC to nearby conducting surfaces because this effect modifies the leakage inductance of the CMC. We have quantified the impact of these effects in some practical cases and we have found that provided that the presence of nearby conducting surfaces is avoided or carefully accounted for, the proposed UDM setup can be used to assess the response of a CMC to DM noise within a range of frequencies of practical interest.

Summing up, the main contribution of the present work is to provide a deep understanding on the actual scope and limitations of a simple technique that can be used to measure the attenuation provided by a common mode choke to differential mode signals. This is important because the simplicity of the proposed measurement technique makes it very appropriate for quick assessment of the suitability of a common mode choke for a particular application. In fact, the quick assessing of the impedance and DM attenuation of a CMC provided by the proposed measurement approach facilitates the tasks of ensuring stability and compliance of electronic equipment with conducted emissions limits imposed by EMC regulations.

Author Contributions: Conceptualization, J.B.-M. and M.A.M.-P.; methodology, J.B.-M.; software, C.D.-P.; validation, P.G.-V. and C.D.-P.; formal analysis, P.G.-V. and C.D.-P.; investigation, P.G.-V. and C.D.-P.; resources, M.A.M.-P.; data curation, C.D.-P.; writing–original draft preparation, J.B.-M. and P.G.-V.; writing–review and editing, J.B.-M. and M.A.M.-P.; visualization, C.D.-P. and P.G.-V.; supervision, M.A.M.-P. and J.B.-M.; project administration, M.A.M.-P.; funding acquisition, J.B.-M. and M.A.M.-P. All authors have read and agreed to the published version of the manuscript.

Funding: This work has been supported by the Spanish Ministerio de Economía y Competitividad (project TEC2014-54097-R) and by H2020-EU.3.4.5.6. SCOPUS project (Grant agreement ID: 831942).

Conflicts of Interest: The authors declare no conflict of interest.

References

1. Kovacevic, I.F.; Friedli, T.; Muesing, A.M.; Kolar, J.W. 3-D electromagnetic modeling of EMI input filters. *IEEE Trans. Ind. Electron.* **2014**, *61*, 231–242. [CrossRef]
2. Paul, C.R. *Introduction to Electromagnetic Compatibility*; John Wiley and Sons: Hoboken, NJ, USA, 2006.
3. Wang, S.; Lee, F.; Chen, D.; Odendaal, W. Effects of parasitic parameters on EMI filter performance. *IEEE Trans. Power Electron.* **2004**, *19*, 869–877. [CrossRef]
4. Wang, S.; Chen, R.; Van Wyk, J.D.; Lee, F.C.; Odendaal, W.G. Developing parasitic cancellation technologies to improve EMI filter performance for switching mode power supplies. *IEEE Trans. Electromagn. Compat.* **2005**, *47*, 921–929. [CrossRef]

5. Chiu, H.J.; Pan, T.F.; Yao, C.J.; Lo, Y.K. Automatic EMI measurement and filter design system for telecom power supplies. *IEEE Trans. Instrum. Meas.* **2007**, *56*, 2254–2261. [CrossRef]
6. Kotny, J.L.; Margueron, X.; Idir, N. High-frequency model of the coupled inductors used in EMI filters. *IEEE Trans. Power Electron.* **2012**, *27*, 2805–2812. [CrossRef]
7. Baccigalupi, A.; Daponte, P.; Grimaldi, D. On a circuit theory approach to evaluate the stray capacitances of two coupled inductors. *IEEE Trans. Instrum. Meas.* **1994**, *43*, 774–776. [CrossRef]
8. Cogitore, B.; Keradec, J.; Barbaroux, J. The two-winding transformer: An experimental method to obtain a wide frequency range equivalent circuit. *IEEE Trans. Instrum. Meas.* **1994**, *43*, 364–371. [CrossRef]
9. Schellmanns, A.; Berrouche, K.; Keradec, J.P. Multiwinding transformers: A successive refinement method to characterize a general equivalent circuit. *IEEE Trans. Instrum. Meas.* **1998**, *47*, 1316–1321. [CrossRef]
10. Margueron, X.; Keradec, J.P. Identifying the magnetic part of the equivalent circuit of n-winding transformers. *IEEE Trans. Instrum. Meas.* **2007**, *56*, 146–152. [CrossRef]
11. Besri, A.; Chazal, H.; Keradec, J.P. Capacitive behavior of HF power transformers: global approach to draw robust equivalent circuits and experimental characterization. In Proceedings of the 2009 IEEE Instrumentation and Measurement Technology Conference, Singapore, 5–7 May 2009; pp. 1262–1267.
12. Roc'h, A.; Bergsma, H.; Zhao, D.; Ferreira, B.; Leferink, F. A new behavioural model for performance evaluation of common mode chokes. In Proceedings of the 2007 18th International Zurich Symposium on Electromagnetic Compatibility, Munich, Germany, 24–28 September 2007; pp. 501–504.
13. Stevanovic, I.; Skibin, S.; Masti, M.; Laitinen, M. Behavioral modeling of chokes for EMI simulations in power electronics. *IEEE Trans. Power Electron.* **2013**, *28*, 695–705. [CrossRef]
14. CISPR 17:2011. Methods of measurement of the suppression characteristics of passive EMC filtering devices. 2011.
15. Kyyrä, J.; Kostov, K. Insertion loss in terms of four-port network parameters. *IET Sci. Meas. Technol.* **2009**, *3*, 208–216.
16. Dominguez-Palacios, C.; Bernal-Méndez, J.; Martin Prats, M.A. Characterization of common mode chokes at high frequencies with simple measurements. *IEEE Trans. Power Electron.* **2018**, *33*, 3975–3987. [CrossRef]
17. den Bossche, A.V.; Valchev, V.C. *Inductors and Transformers for Power Electronics*; CRC Press: Boca Raton, FL, USA, 2005.
18. Bockelman, D.E.; Eisenstadt, W.R. Combined differential and common-mode scattering parameters: Theory and simulation. *IEEE Trans. Microw. Theory Technol.* **1995**, *43*, 1530–1539. [CrossRef]
19. Bernal-Méndez, J.; Freire, M.J.; Martin Prats, M.A. Overcoming the effect of test fixtures on the measurement of parasitics of capacitors and inductors. *IEEE Trans. Power Electron.* **2020**, *35*, 15–19. [CrossRef]
20. Pozar, D. *Microwave Engineering*; Wiley&Son: Hoboken, NJ, USA, 2011.
21. Nave, M. *Power Line Filter Design for Switched-Mode Power Supplies*; Van Nostrand Reinhold: New York, NY, USA, 1991.
22. Dominguez Palacios, C.; Gonzalez Vizuete, P.; Martin Prats, M.A.; Bernal, J. Smart shielding techniques for common mode chokes in EMI filters. *IEEE Trans. Electromagn. Compat.* **2019**, *61*, 1329–1336. [CrossRef]
23. Bernal-Méndez, J.; Freire, M.J. On-site, quick and cost-effective techniques for improving the performance of EMI filters by using conducting bands. In Proceedings of the 2016 IEEE International Symposium on Electromagnetic Compatibility, Ottawa, ON, Canada, 25–29 July 2016; pp. 390–395.
24. Weile, D.; Michielssen, E. Genetic algorithm optimization applied to electromagnetics: A review. *IEEE Trans. Antennas Propag.* **1997**, *45*, 343–353. [CrossRef]
25. EN55011:2011/CISPR 11. *Industrial, Scientific, and Medical Equipment—Radio-Frequency Disturbance Characteristics—Limits and Methods of Measurement*; Beuth Verlag: Berlin, Germany, 2009.
26. EN55022:2011/CISPR 22. *Information Technology Equipment—Radio disturbance characteristics—Limits and methods of measurement*; Beuth Verlag: Berlin, Germany, 2008.
27. FCC Part 15. *The FCC 47 CFR Part 15 from the Federal Communications Commission: Rules and Regulations for EMC*; Federal Communications Commission: Washington, DC, USA, 2020.
28. Available online: https://www.coilcraft.com/wbt.cfm (accessed on 15 March 2012).

MDPI
St. Alban-Anlage 66
4052 Basel
Switzerland
Tel. +41 61 683 77 34
Fax +41 61 302 89 18
www.mdpi.com

Electronics Editorial Office
E-mail: electronics@mdpi.com
www.mdpi.com/journal/electronics

www.ingramcontent.com/pod-product-compliance
Lightning Source LLC
LaVergne TN
LVHW071008170726
843515LV00006B/1112

* 9 7 8 3 0 3 6 5 0 5 0 0 8 *